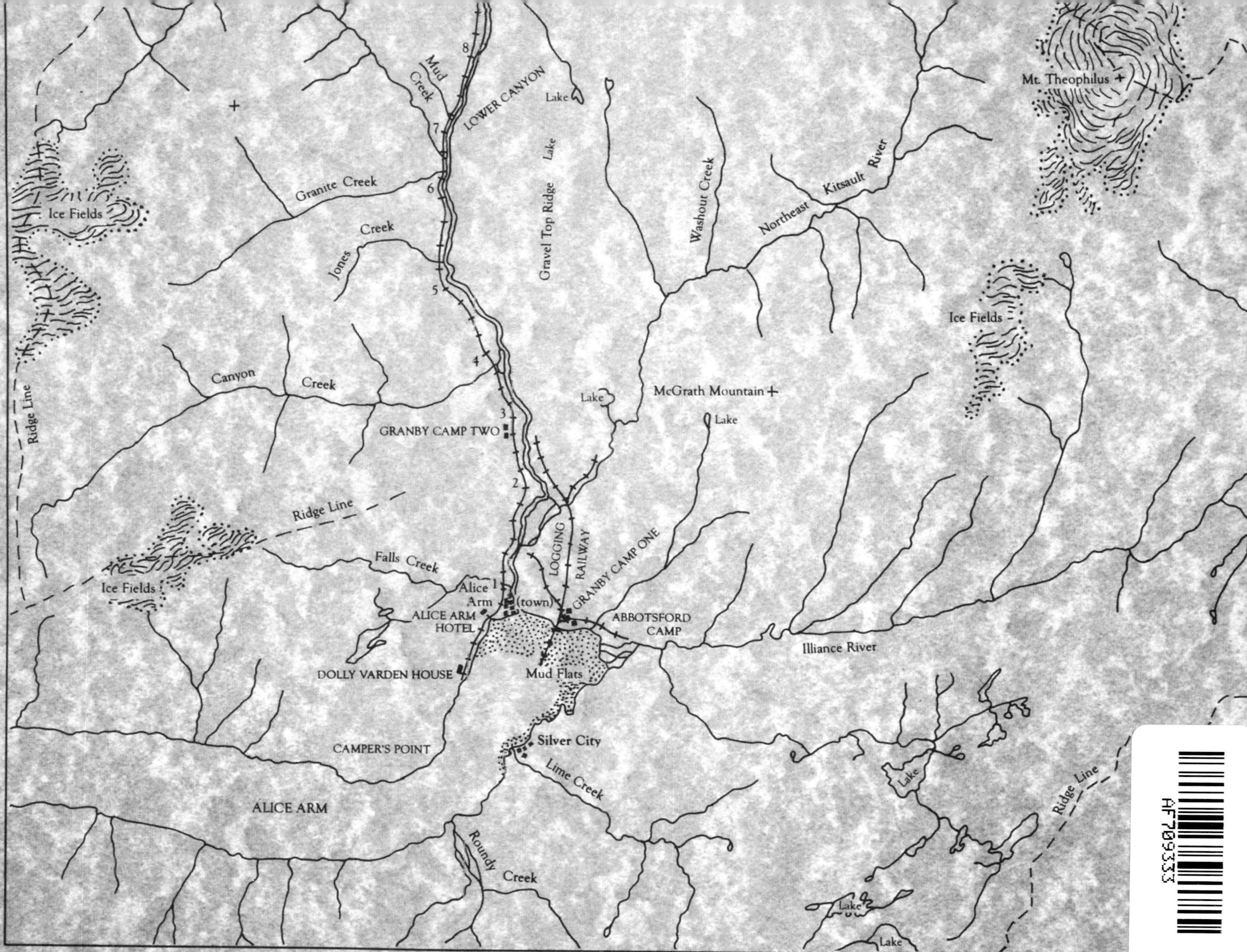

Mt. Theophilus
Mud Creek
LOWER CANYON
Lake
Gravel Top Ridge
Lake
Washout Creek
Northeast Kitsault River
Granite Creek
Ice Fields
Jones Creek
Ice Fields
Canyon Creek
McGrath Mountain
Lake
Lake
GRANBY CAMP TWO
Ridge Line
Ridge Line
LOGGING RAILWAY
GRANBY CAMP ONE
Falls Creek
Alice Arm
(town)
ALICE ARM HOTEL
ABBOTSFORD CAMP
Illiance River
Ice Fields
DOLLY VARDEN HOUSE
Mud Flats
Silver City
CAMPER'S POINT
Lime Creek
Lake
Ridge Line
ALICE ARM
Roundy Creek
Lake
Lake
1
2
3
4
5
6
7
8

STEEL RAILS & SILVER DREAMS

A HISTORY OF THE DOLLY VARDEN MINES AND THE NARROW GAUGE DOLLY VARDEN MINES RAILWAY

By DARRYL E. MURALT

With Notes on the Railway Operations and Equipment of the Granby Consolidated Mining, Smelting and Power Company at Anyox, B.C. by John Lewis.

Library of Congress Cataloging-in-Publication Data

Muralt, Darryl E., 1944-
Steel rails & silver dreams.

Bibliography: p.
Includes index.
1. Dolly Varden Mines. 2. Dolly Varden Mines Railway. 3. Silver mines and mining - British Columbia.
1. Title. II. Title: Streel Rails and silver dreams.
TN439.D65M87 1985 385'.54'0971132 85-18718

ISBN 0-9615467-0-0

Second printing 2007
Printed in China on recycled acid free paper

Dedication

To the memory of
Ole Evindsen, 1891 – 1971
and to the memories of
Ole Pearson, K. L. Eik and
E. Carlson; friends, dreamers
and partners in an
incredible quest.

Painting by Bob Schlechter

Steel Rails and Silver Dreams

It's a place of steel rails and silver dreams
And miner's wildest get-rich schemes come true
It's a place with wild rivers raging
And glaciers cold and icy blue
It's a place where the snows lie deep in winter
And the rains fall hard in spring
Where the land is ruled by the grizzly
And in the sky the eagle's king
It's a place of steel rails and silver dreams
And man's unbroken will
And a silver hoard's still waiting there
At the Dolly Varden Hill

Darryl E. Muralt
August 1984

Preface

This is the story of the Dolly Varden Mine and the Dolly Varden Mines Railway, 18 miles of improbable narrow gauge iron reaching north from the little community of Alice Arm to this rich silver property, now a legend in the annals of mining in British Columbia. In the years between 1916 and 1922, the Dolly Varden Mine was both the brightest hope and the greatest disappointment of the mining industry of Northwest Canada.

The mine contained pockets of ore that assayed as high as 2200 ounces of silver per ton. But, in spite of its great wealth and the best efforts of some of the most talented mining men of the day, it was beset by legal and financial problems that led to its closure by the summer of 1921. Today, it is amalgamated with several other properties as Dolly Varden Minerals Inc. Together these properties may contain the largest remaining reserve of silver in North America.

The Dolly Varden Mine was discovered by four Scandinavian prospectors in 1910 just as foretold in a mysterious dream. However, as has so often been the case, its co-discoverers were not to be the primary beneficiaries of its development. In 1915 the mine was sold to a group of Chicago capitalists who incorporated the Dolly Varden Mines Company and undertook to bring it into production.

The site of the Dolly Varden Mine is in rugged mountain terrain, 18 miles north of Alice Arm. A railway was needed to transport equipment and supplies up to the mine and to ship the silver ore down to tidewater for transfer to the smelter. The railway construction contract was let to the Taylor Engineering Company of Vancouver, B. C., in February of 1917. Construction began in the summer of that year and was far more difficult than anticipated, continuing into the following year.

The Dolly Varden Mines Railway was the last narrow gauge common carrier built in British Columbia. As such it belonged to a select group of little railways that include the Trail Creek Tramway, the Columbia and Western Railway, the Kaslo and Slocan, and the White Pass & Yukon Route. Narrow gauge didn't enjoy the same popularity in Canada as it did in the United States. In spite of this, its economy of construction made it a natural for building railways into otherwise inaccessible mountain areas such as the site of the Dolly Varden Mine.

Differences arose between the Dolly Varden Mines Company and the Taylor Engineering Company that led to the bankruptcy of the Taylor Company in November of 1918. Because of legal technicalities, these differences could not be referred to the courts. The two parties appealed to the government of British Columbia for a solution and the matter was finally referred to a select committee of the Provincial Legislature. A hearing was held that resulted in the mine and railroad being turned over to the Taylor Engineering Company, on the condition that the Taylor Engineering Company would reimburse the Dolly Varden Company for its investment in these properties.

The Taylor Engineering Company (reorganized as the Taylor Mining Company) took over the mine and railroad in June of 1919; the railway was rushed to completion within two months and ore shipments were begun. In the three years 1919, 1920, 1921, the Dolly Varden Mine shipped over 36,000 tons of ore and could have gone on shipping more in later years had economic conditions warranted it. Unfortunately, a depression in the silver market and the lack of sufficient development capital forced the Taylor Mining Company to lose control of the properties in 1921-22. Because of the huge debt assumed at the time of its takeover of the mine and railway and the high costs involved in bringing the mine into production, the Taylor Mining Company was unable to meet its financial obligations or to secure additional operating capital. The mine and railroad were taken over by a creditor and later sold to the David Spencer interests of Vancouver.

Here then, is the story of the spectacular Dolly Varden and the little narrow gauge railroad that connected it with the outside world – a story of compelling interest to aficionados of British Columbia history as well as those who follow the chronicles of mining and the iron horse. The Dolly Varden story has silver, scandal, and steel rails – enough to excite the most complacent admirer of a time not so long ago when British Columbia was a much younger province in the bloom of its pioneering youth.

Table of Contents

Maps, Illustrations, and Plans

Note: *In this book, the north indicated on maps is true north, not magnetic north.*

Tables

Acknowledgements

No book ever comes to be without the assistance and advice of many people other than the author. *Steel Rails and Silver Dreams* is no exception to that rule and I have many people to thank. They include the staff of the Provincial Legislative Library in Victoria and in particular, William Peters, who located the original transcripts of the Dolly Varden Hearing that were so valuable.

A very special thanks to Ronald Evindsen and Alma Graham, the son and daughter of Ole Evindsen, who shared many precious family memories and permitted the use of Ole's photo collection. Early assistance and inspiration also came from Drs. Ted Grove and Nick Carter, and Fraser McLean who shared their knowledge of the geology and mines of the Alice Arm-Observatory Inlet area and from Mr. Harry Bapty, the former resident manager of the Torbrit Mines.

One of the most gratifying aspects of preparing this book was the wonderful "hands across the border" kindness of Richard Dunn who is responsible for the superb drawings of the locomotives and cars. Thanks also to Ron Anstey for his drawing.

John Lewis, who has assisted with the equipment rosters and contributed the Appendix on the Granby mining railways is a railroad author and recognized authority on narrow gauge logging and mining railways. I am extremely grateful for his kindness and patience over the past few years.

David Wilkie, Elwood White and R. Ken Bradley of the British Columbia Railway Historical Association kindly gave critical readings. Ted Wrinkle gave valuable editorial advice and assistance during the early stages of the project.

In 1979 I was fortunate enough to visit the town of Alice Arm and I make special mention of the kindness of Inge Fiva, a long time resident of Alice Arm and one of the last of a rare breed, the old-time prospector. I wish for his sake that this could have been a history of the town of Alice Arm, but time and distance have dictated that the book restrict itself to the story of the Dolly Varden Mine and the Railway. Another helpful former resident of the town is Mr. C. Marshall Smith who contributed advice and many reminiscences.

More recently, Douglas Richter, an expert on locomotive history and geneology has contributed much valuable information, photos and inspiration without which this book would have been much less informative. Another expert who gave advice and photographic assistance was Mr. Walt Casler of Corry, Pennsylvania, a former engineer for the Climax Locomotive Company who spent much of his early career delivering and setting up locomotives and who is well known in more recent years for his history of the Climax Locomotive.

I was particularly pleased when Robert D. (Bob) Schlechter, a recognized railroad artist and draftsman consented to do the color painting used as a frontispiece for the book.

Bob and Irene Brown of Benchmark Publications Ltd., the publishers of this book, my first major work, deserve special recognition and thanks for their confidence in the story and their dedication to seeing it in print. Thanks are also due Sharon Olsen who designed the book, John Bird who typeset it, and Barbara Marsh who copyedited it.

To all of these people and to any of the many others whose names are not mentioned because of the dictates of space, I thank you all for helping me realize the dream of making the legend of the Dolly Varden live again.

Darryl E. Muralt
June 1984

3272
DOMINION
PHOTO CO.
VANCOUVER B.C.

1

Early History and Development

To some, the name Dolly Varden brings to mind the young heroine of Charles Dickens' story of Barnaby Rudge. To others, it evokes memories of the scrappy little Dolly Varden trout. But to a dwindling number of British Columbians, the name Dolly Varden revives memories of a bonanza of shining silver locked in the canyons and hills of the Kitsault River Valley near the little town of Alice Arm.

The town of Alice Arm is located on the northwest coast of British Columbia, Canada, at the head of a narrow inlet, also called Alice Arm. This inlet is 18 miles long and branches east and then north from Observatory Inlet near the British Columbia-Alaska border. It is a narrow, dark stretch of water lying between steep-sided hills with dense evergreen forests stretching up from the water's edge to rounded heights.

The town and the inlet are named for Alice, wife of the Reverend Robert Tomlinson. Tomlinson had charge of the Anglican Church Missionary Society's mission at Kincolith on the Nass River. In the 1860s, Alice Tomlinson put behind the comforts of civilization in early-day Victoria to join her husband in his mission to the native peoples of the northwest coast. The trip up to Kincolith from Victoria was made in an open dugout canoe and took several weeks. In later years, she traveled widely with her husband on numerous other expeditions of similar duration. The inlet of Alice Arm was named in her honor in 1869. When I was preparing to write this book, I stood on the foreshore at the head of Alice Arm and wondered what Alice Tomlinson's vision for the future of this remote part of the province might have been. Could she have ever imagined or foreseen the drama that unfolded there within a generation of her passing, when the discovery of rich deposits of silver, copper, gold, and molybdenum led to the founding of the town named also in her honor?

In centuries past, the area that became Alice Arm was the site of an Indian encampment. The Nishga people stayed there when they crossed over the mountains from the Nass River area, apparently to hunt mountain goats on the surrounding peaks. These brief seasonal occupations had ended by the turn of the century and today there is little evidence of a permanent Indian settlement in the area. Nevertheless, an Indian reservation existed in the area until 1912, when the land it occupied was transferred from the Canadian federal government to the provincial government of British Columbia. In 1929, this land was finally surveyed into town lots and sold at public auction; the moneys from the sale were divided between the Province of British Columbia and the Canadian Department of Indian Affairs.

The first white men known to have visited the Alice Arm area were the crew of the HMS *Resolution*, under the command of Capt. George Vancouver, who explored and named Observatory Inlet during a hydrographic expedition in 1793. In 1874, the

Left:
The Kitsault River in the Mud Creek Canyon above Mile 7 at low water. The railway was susceptible to washouts here. During construction in 1917, men had to be lowered down to the cliff face on ropes to carve out the first perilous footholds for the railway.

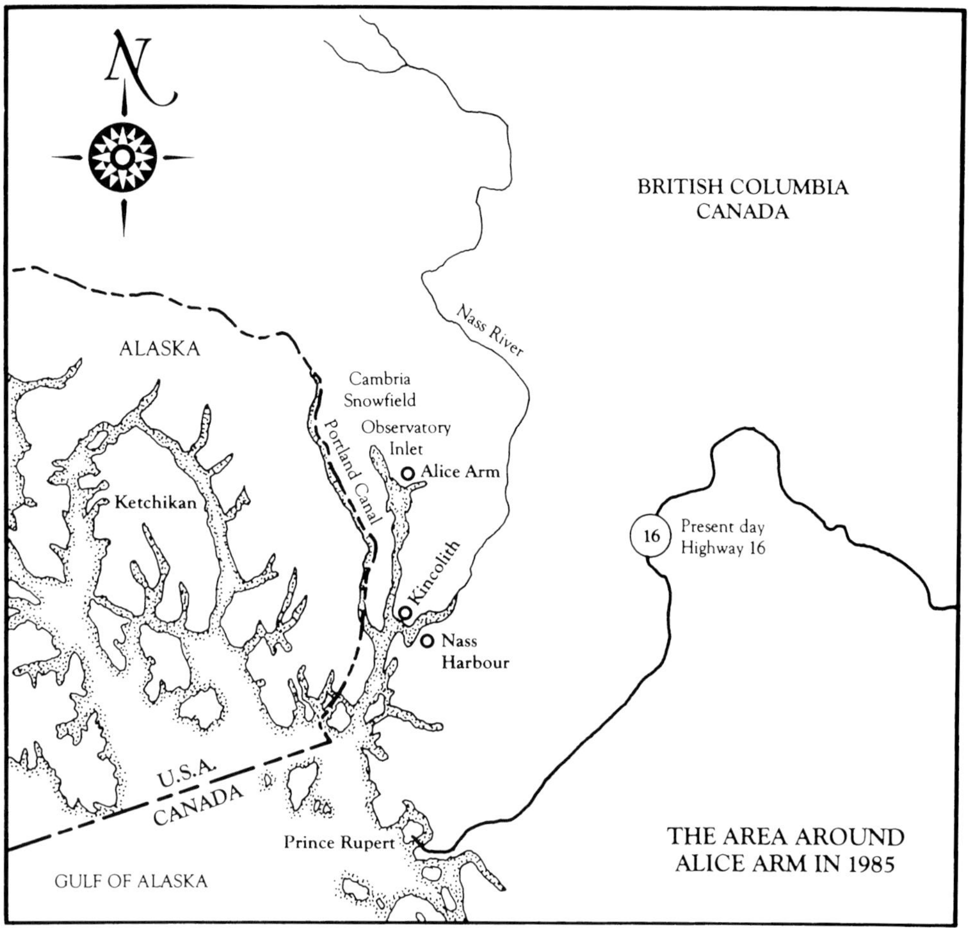

THE AREA AROUND ALICE ARM IN 1985

The Western coast of North America, showing Alice Arm.

Lyall Expedition was commissioned to explore the valley stretching north from Alice Arm in order to find a low pass over the mountains from the coast into the Cassiar country, an important mining and fur trading area. The Lyall Party explored the valley as far north as the Kitsault Glacier and found the low pass; however, the route was unsuitable for horses because it lacked forage. This was the first recorded passage of white men through this area that had known only the passing of the seasons and an occasional native hunting party through the long centuries following the valley's creation during the last Ice Age. The Lyall party uncovered a coal seam from which they extracted fuel for their cooking fires, but they passed over, without notice, the rich metalliferous zones that were to bring the valley to the attention of the mining world a half-century later. In like manner, more recent explorers have passed over the coal seam without noticing it; searches for it have been unsuccessful.

The Alice Arm district contains the valleys of the Kitsault and the Northeast Kitsault Rivers, tributaries to Alice Arm from the north, and the valley of the Illiance River, a tributary from the northeast. The district extends down the inlet to the area beyond the new mining town of Kitsault (formerly Silver City). The main Kitsault Valley stretches north from the town of Alice Arm for 26 miles to the Kitsault Glacier where the Kitsault River is born among the myriad streamlets at its icy base. The low pass discovered by the Lyall Expedition is filled with the ice fields of this glacier that block the way to the Nass River country and the Cassiar country beyond.

The Kitsault Valley is U-shaped and steep-sided with hills rising to about 3000 feet before tapering off to rounded heights. Beyond these hills are peaks ranging from 4000 to 6000 feet; many are snow covered and flanked by permanent ice fields. The main valley is open in its lower reaches but narrows into a steep-walled gorge where the icy waters boil around jagged rocks during the spring run-off and during the frequent storms that precede the winter season. In some places the lower level of the canyon is impassable and trails (and later roads) had to be built on the heights. This gorge stretches from Miner Creek, 14½ miles north of the town, to Evindsen Creek which enters the Kitsault from the west at a point 18 miles up. There is a smaller defile known as the Lower or Mud Creek Canyon at a point about 7 miles up-river from Alice Arm.

The Kitsault's tributary valleys are flat-bottomed with creeks that are flat near their mouths but steep in their upper reaches. The upstream part of the area is heavily wooded; stands of hemlock and spruce extend up to the sub-alpine zone at about the 3000-foot elevation. Downstream, large stands of cottonwoods grace the deltas of the Kitsault and Illiance Rivers.

Annual precipitation in the Alice Arm area normally aggregates in excess of 100 inches. The upper valley is impassable from late October to May during the heavy winter snows that are frequently 20 or more feet deep. Summer mornings are cool on Alice Arm and dew-like droplets of moisture hang heavy in the air, drifting across the harbor before burning off in the late morning sun.

THE PROSPECTORS ARRIVE

No one knows exactly when the first mineral claims were staked at Alice Arm. Before the turn of the century the Nishga Indians had shown an Anglican missionary, Archdeacon Collinson, deposits of molybdenum near what is now the mining town of Kitsault. These deposits were examined by British mining engineers but no claims are recorded from that early time. Today, however, these deposits are the site of the huge molybdenum mine of the AMAX Corporation.

Frank Roundy, an experienced prospector, first visited Alice Arm in 1888. In 1903, he staked three claims close to what was to become the town of Alice Arm: the I'll Chance It, the Black Bear, and the Aldebaran. The known presence of precious metal on his claims, all located close to tidewater, led to the establishment of a small settlement.

The first permanent resident of the area was John Stark, who arrived in 1901. Stark was a member of a prominent Negro family that had come north from the United States to settle on Salt Spring Island, near Victoria, some years earlier. In his later years, Stark often expressed pride at having been the first "white man" to reside in the Alice Arm area. Stark staked two claims, the Wolf (not to be confused with the other Wolf staked later near the Dolly Varden) next to Roundy's claims, and the Bellevue, in the Illiance River Valley.

With the successful staking of these claims, other prospectors came in to investigate the area. In the years between 1907 and 1910, a number of other finds were made. The molybdenum deposits first shown to Archdeacon Collinson were re-discovered and stakings were made about 4 miles down the Arm in 1907. In 1908, Frank Jones and Benny Williams staked claims on the western slope of McGrath Mountain between the Kitsault and Illiance Rivers. In the same year, John Strombeck and George Young staked the Alice and the Silver Ridge claims. Also in 1908, Fred Juggins located two claims on what became known as the "Red Point Property" (not to be confused with the Red Point claim later staked near the Dolly Varden). Several other claims were staked on the west side of the harbor.

In 1908 Joe Wells and William Dilworth ascended the Kitsault Valley and crossed over the divide into the Nass River watershed. They were the first white men to make this trip. In the following year they explored the northeast fork of the Kitsault River and staked the claims they called Emerald and Albion. Thus far, all claims including these had been only 3 or 4 miles from the harbor; there had been little prospecting further inland than that because of the dense forest and the rugged terrain. In fact, before 1910, no one except the natives had explored the middle and upper reaches of the Kitsault and its secrets remained hidden. In that year, the arrival of four Scandinavian prospectors on the coastal steamer from Prince Rupert changed all that and led to the discovery of large deposits of silver-and copper-bearing ores in the upper Kitsault Valley.

Above:
John Stark, left front, mugging for the camera, has a drink from a bottle of spirits held by a Mrs. Thompson. The man in the fedora is Jack Anderson, a prospector. In the back row, left to right, are Mrs. Wheatly, Mrs. Peterson and Mrs. McKinley, wives of local residents. Stark, a Negro, was proud of having been the first "white man" to reside in the Alice Arm area. This snapshot was taken by Ole Evindsen.

Below:
Before 1910, mining claims in the Alice Arm area were all within 3 or 4 miles from the harbor.This view taken by Ole Evindsen in 1915 from Silver City (now the mining town of Kitsault) shows the rugged terrain that made prospecting difficult futher inland. Alice Arm is in the foreground and Mount Theophilus is on the left.

STAKING THE DREAM MINE

Early in July of 1910, Ole Evindsen, Ole Pearson, K. L. Eik, and E. Carlson arrived at Alice Arm on the weekly steamer with the intention of exploring the middle and upper Kitsault Valley. Although these men have often been described as "four Swedish prospectors," Evindsen was born in northern Norway. Perhaps the confusion arose because he spoke fluent Swedish. In 1920, Alfred Taylor who had by then become a central figure in the Dolly Varden story, listened and recorded Evindsen's tale of the expedition for the *Mining and Scientific Press* of May, 1920 (1.1). It's not hard to imagine Evindsen spinning his tale, sitting in the smoke-filled bar with Taylor and the circle of regulars consisting mainly of grizzled, bush-wise prospectors who were in from the hills until spring:

"I heard of mineral wealth so I left Prince Rupert and came up here with three other chaps in the spring of 1910. We packed in grub and set out up the Kitsault River. None of us had been there before but we hoped to find some good claims. One night just before we started, one of the chaps, Ole Pearson, had a dream.

"His uncle had died a few months before. Ole told us that his uncle had come to him in a dream the night before and showed him a big white boulder and said, 'You are going to find a big mine which will make you rich so that you will not need to work for the rest of your life. You will go on until you find this boulder and these rocks. Then you must stake the claim and the name you must give it is the Dolly Varden . . . '"

To shorten the story a bit, Ole Pearson told his friends about the dream and they were amused. However, it seemed a fortuitous sign and the small party headed up the valley with high hopes. For several days, they climbed through the trackless canyon. Then Eik and Carlson headed back down to Alice Arm for supplies while Evindsen, the most experienced prospector, and Pearson, the dreamer, continued on. Thus far, there had been no sign of the bonanza ore. We may reasonably assume that the party had been passing over the tufaceous zone along both sides of the Kitsault River between the mouth of the northeast fork of the Kitsault River and Mud Creek Canyon. Now let Evindsen continue:

Ludwig Eik, one of the original four partners, on a visit to Alice Arm from his Saskatchewan home in the late 1930s.

"We took different paths about 75 yards apart. Then I saw in the bare andesite a streak of grey rock that looked like it carried silver. With my hammer I broke down the capping of the rock and saw I'd discovered a vein that looked like rich ore.

"While I was examining my find, Ole Pearson let out a yell. 'I've found the mine! I've found the mine!'"

Pearson had apparently discovered the white boulder described to him in his dream. He had not yet sighted any mineral. Evindsen ran down the hill with his samples and the two were ecstatic over their discovery. Between the silver found by Evindsen and the large white boulder of Pearson's dream, they were both sure the vision had foretold reality.

"'Say, what was the name your uncle said we were to call the claim?' I asked. 'Hanged if I can remember,' admitted Ole Pearson at last, after trying to cudgel his brains.'"

And to shorten the story at bit again, for 2 days Evindsen and Pearson sat trying desperately to remember, but to no avail. Little wonder they could not remember, "Dolly Varden" was the name of a character from Charles Dickens' *Barnaby Rudge* and hardly a name you'd expect a Swedish (or Norwegian) prospector to have on the tip of his tongue. Evindsen and Pearson passed 2 trying days before their partners finally came in sight from down-river. When they had come within hailing distance, Evindsen quickly popped the fatal question:

"'Say, what was the name of the claim that Ole Pearson's uncle told him about?' 'It was the Dolly Varden,' came the reply."

And so the mine was named and years later Ole Evindsen would often recite the tale of the mine's discovery while he passed an evening with the guests at his snug hostelry by the mouth of the Kitsault. How fortunate it was that one of the party had been able to remember the name, Dolly Varden. Otherwise, who knows, perhaps the mine might have proved as ethereal as Pearson's visit from his departed uncle.

Ole Pearson (left) and Ole Evindsen (right), a few years before their discovery of the "Dream Mine."

The Dolly Varden properties were staked on the steep west bank of the Kitsault River at an altitude of 2076 feet. The location was on a mountain on the southeast side of Evindsen Creek and between that stream and the Kitsault River. The claims were just over 17 miles up-river from Alice Arm. Part of the area was covered with poor quality timber choked with deadfalls. There was a layer of glacial till that made it difficult to determine the nature of the rocks lying beneath, and snowfall in the area often reached 20 feet. All this would make development of the Dolly Varden Mine slow, expensive, and extremely difficult. And however rich its ores, the Dolly Varden could not be properly exploited until a transportation system to move men, equipment, and supplies in and ores out had been created. By 1913, a trail about 22 miles in length led into the area, but it was impassable to horses beyond a point about 7

Ole Evindsen took this photo of a footbridge over Evindsen Creek at its confluence with the Kitsault River about 18 miles from Alice Arm. Prospecting was difficult in this area because of dense forest and a layer of glacial till that obscured the underlying rocks.

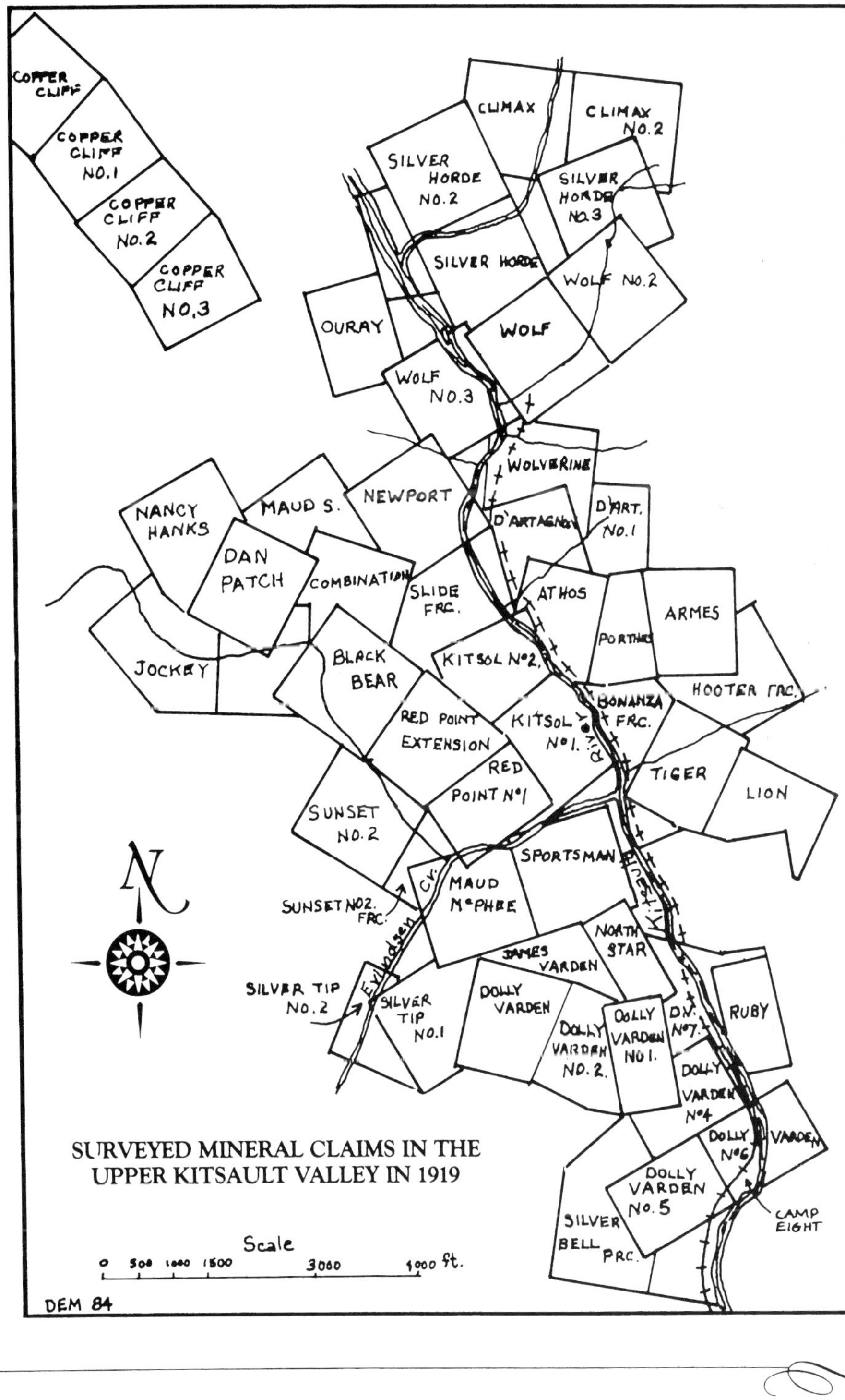

SURVEYED MINERAL CLAIMS IN THE UPPER KITSAULT VALLEY IN 1919

miles up the valley from the harbor; beyond that point, all supplies and equipment had to be carried in on the backs of the prospectors.

The Dolly Varden consisted of eight claims covering an area of 213 acres. They extended from the bank of the Kitsault to the top of what was later called Dolly Varden Hill. The first prospecting camp was located not far above the river level but the surface buildings and main workings of the mine were eventually located at an elevation of about 1700 feet, some 700 feet above the water level of the river.

The ore bodies were classified as replacement deposits and occurred along fracture zones in the andesite, a volcanic rock. As the volcanic material cooled, it cracked and the ores were deposited by water flowing through the cracks over long periods of time. There were significant bodies of quartz that were accompanied by pyrite, galena, sphalerite, native silver, ruby silver, argentite, and rhodochrosite.

EARLY DEVELOPMENTS AT THE DOLLY VARDEN

As has so often been the case with the discovery of precious metals, the locators were not to be the major beneficiaries of the Dolly Varden's discovery and development. Although the four prospectors filed their claims in 1910, it was to be several years before there were any tangible returns from their discovery. Nevertheless, Evindsen and his partners did hold the properties for a time and carried out exploration and development work. The four prospectors gave one-fifth interest in the mine to Charlie Swanson in exchange for a grubstake. With only their own resources and the Swanson money for support, they began an attempt to develop the mine. They constructed a cabin and began digging a tunnel.

However, they were distracted from their work at the Dolly Varden by the need to protect their interests in the area. Their discovery at the Dolly Varden had attracted the attention of other prospectors and a number of claims were being staked in the upper Kitsault River area. The four Scandinavians set aside work on the "Dolly" for a time while they staked several other properties. The most promising of these was the Red Point which was staked just a short time after the Dolly Varden. It was a copper property, located along the bluffs on the north side of Evindsen Creek almost opposite the Dolly Varden.

When D. G. Forbes of the British Columbia Department of Mines visited the upper Kitsault area in 1912, he examined the Red Point, noting that: "At an elevation of 2025 feet an exposure of ore has been laid bare at the top of the bluff, from which sample no. 1 was taken as an average across the face and assayed: copper, 8.5%; silver, 1.3 oz.; gold, $7.60." (1.2)

The Dolly Varden was mentioned only briefly in his report: "A cut of 45 feet has been made across a large body of quartz striking east and west. The cut does not expose the walls of the deposit. The quartz contains pyrite, galena, arsenopyrite, and chalcopyrite. A general sample assayed: gold, 0.02 oz.; silver 1.4 oz.; copper, 1.2%. This was the only opening on the claim seen by the writer, but another is said to exist at a lower level showing better values." (1.3)

Compared to assays from the Red Point, the single sample from the Dolly Varden was of little interest, with the result that no one other than the owners was likely to be very excited about the Dolly Varden property.

Around 1912, Evindsen and his partners "bonded" (a bond was the jargon of the time for what today is called a long-term lease) the Red Point to A. D. McPhee of Seattle who "re-bonded" it to a Mr. Hopkins. It was Hopkins who, in 1915, persuaded the engineer, R. B. McGinnis of San Francisco (representing a group of investors in Chicago) to examine the Red Point. During his examination of the copper property, McGinnis was shown a sample of high-grade ore from lower down on the Dolly Varden, the site which had not been visited by D. G. Forbes of the Department of Mines during his visit. McGinnis was so impressed that he forgot about the Red Point mine site and immediately "bonded" the Dolly Varden for a sum of $50,000.00. At

Ole Evindsen took this photo of the four Scandinavians' fifth partner, Charlie Swanson. In 1912, Swanson traded a grubstake for a one-fifth interest in the Dolly Varden Mine.

Ole Evindsen with his lead dog, Sport, before a trip up to the Dolly Varden Mine.

Ole Evindsen transporting supplies to the Dolly Varden Mine by dog sled in the winter of 1914.

Smoke billows from the chimney of the first Dolly Varden Mine camp during the deep winter of 1914.

The Dolly Varden crew pose for the camera in 1915 with visiting engineer, R. B. McGinnis. When McGinnis saw the rich Dolly Varden ore, he forked over $50,000.00 for a long-term lease on the mine. Hiring Ole and his crew to continue developing the mine, he contacted the Chicago financiers he represented for further funding. Mine workers, left to right: a Mr. Frizzell, W. J. Crawford, George Salt, discoverer Ole Evindsen; and engineer R. B. McGinnis.

Ole and Svea Evindsen on their wedding day, May 8, 1915.

Left:
Ole Evindsen took this photo of his Alice Arm Hotel when it was under construction in 1916. He used the money he received from R. B. McGinnis for his interest in the Dolly Varden Mine to build the hotel on a hill overlooking the harbor front.

Svea Evindsen helped her husband by working at the Dolly Varden Mine. Ole photographed her with her horse ready for a ride up to the mine in 1915.

Ole and Svea Evindsen with their dog, Sport, on the trail to the Dolly Varden in the summer of 1915.

Ole Evindsen photographed this evening gathering in his cabin at the Dolly Varden Mine in 1915. Left to right, an unidentified miner; Andrew Hanson (Svea Evindsen's father); Svea Evindsen; miner, Oscar Flint, an unknown miner; miner Gus Anderson. Flint and Anderson were hired by Evindsen to help in the mine.

last, the four Scandinavians got some return on their discovery of the dream mine. The original four partners received $10,000.00 each. Swanson received $10,000.00 for the one-fifth interest he had obtained in exchange for the grub stake.

Pearson's share of the money enabled him to realize at least part of his dream of 5 years earlier: he returned to his native country and bought a farm. Carlson, Eik, Evindsen, and Swanson continued to work on the mine for awhile under McGinnis' direction. Carlson stayed in Alice Arm until his death at the Anyox Hospital during the influenza epidemic of 1918. Eik eventually moved to Saskatchewan. The fate of Swanson is unknown.

Evindsen remained in Alice Arm and became one of its leading citizens, eventually using his share of the money to build his Alice Arm Hotel in 1916. More immediately, he married Svea Hanson on May 8, 1915, and Svea and her father became part of the team of workers attempting to develop the mine. Evindsen hired a local widow, Mrs. Handley, to cook for the Dolly Varden camp and, to judge from the photos, much hard work and many pleasant times were part of this early phase of development.

Meanwhile, R. B. McGinnis had "re-bonded" the Dolly Varden claims to the Chicago interests he represented, and it remained to be seen what actions would be taken by the Chicago investors to develop the mine to the production stage.

3303
DOMINION
PHOTO CO.

2

Opening Moves: 1915 – 1916

FORMATION OF THE DOLLY VARDEN MINES COMPANY

The Dolly Varden Mines Company was formed in Chicago in 1915 and incorporated in Delaware. It was capitalized at $1,250,000.00 consisting of $500,000.00 in preferred stock and $750,000.00 in common shares at a par value of $100.00 per share. Joseph Hubbard, an experienced mining entrepreneur, was made president. Hubbard held directorships on a number of corporate boards and was vice president of the Goldfield Consolidated Mines Company of Reno, Nevada. His role as president of the Dolly Varden Company was to invest the stockholders' funds and to manage the actual collection and disbursement of moneys.

The Dolly Varden shareholders were a group of about 20 wealthy people who pooled their moneys to invest in promising mining properties. Some of them held influential positions in both government and business and were called "One Dollar Men" because of the token one-dollar payment they received for their participation on the boards of directors of various firms. One of the Dolly Varden stockholders was Herbert Hoover who later became President of the United States.

The Dolly Varden Company was made up of a group of people who were playing an investment game. They were prepared to lose a little in speculation from time to time, but this time Hubbard's reputation led them to believe the odds were in favor of their success. In spite of the general acceptability of an occasional loss, some of the stockholders were concerned that several of their recent investments had cost more than anticipated and they resolved to limit the amount of money they were prepared to spend to open the Dolly Varden Mine.

WORK AT THE DOLLY VARDEN, 1915 – 1916

Development work continued at the mine under the direction of R. B. McGinnis and progressed well during 1915. McGinnis had hired Evindsen and his crew and their efforts were directed at "proving-up" or establishing the size of the Dolly Varden ore body. Several large open cuts were excavated on the surface and 550 feet of tunnels were driven into the upper part of the Dolly Varden Hill. Forty-five hundred feet of diamond drilling reached out in all directions and showed that the ore body covered a wider area and reached to a greater depth than had originally been thought. Brittle silver ore, ruby silver ore, and native silver were found down to a depth of over 600 feet. The ore was concentrated in several large masses that had been created by faulting or lateral displacement of the original rock formation. The work of the summer and fall of 1915 was so successful that it was carried on into the winter months with two shifts of miners.

Left:
The grade of the Dolly Varden Mines Railway just north of Camp 8 on the way to the Wolf Mine.

Ole Evindsen photographed Mrs. Handley, the Dolly Varden camp cook, with her two children in front of the cook tent in 1915. Mrs. Handley, a widow from Alice Arm, was one of the original workers hired by Evindsen when he was attempting to develop the mine before it was leased to the Dolly Varden Company.

Early in 1916 the Dolly Varden Mines Company, Limited was granted permission to operate in British Columbia as an extra-provincial company. While this formality was being attended to, McGinnis moved steadily ahead with the exploration of the Dolly Varden ore body. The tonnage of ore "proved-out" was sufficient to justify large volume shipments that would need an adequate transportation system.

In the spring of 1916 the Dolly Varden Mine Company allocated $40,000.00 to improve the prospector's trail as far as the mine; this trail ran up the west bank of the Kitsault River from Alice Arm, past the mine to the head of the Kitsault Valley. Building a trail was extremely difficult in this rugged country. Trees, brush, and deadfalls had to be removed before grading could begin. Hand tools and dynamite were used to widen the trail for pack horses. The hand work was complicated by alternating stretches of rock and swamp. The swamps could only be crossed by a corduroy road made from logs cut along the trail. These logs were laid down into the mire in layers with each layer at right angles to the last layer, until the mat of logs was solid enough to support the loaded pack horses. In many places, rocks had to be blasted to provide clearance for the 6-foot-wide trail.

Snow in the late spring, rain and flash-floods in the summer and fall, and hordes of mosquitoes, blackflies and other stinging insects made construction difficult. Remoteness from population centers and the absence of young men (because of World War I) made it extremely difficult to hire skilled or unskilled laborers. But in spite of all of these problems McGinnis and his crew were able to make better progress than he had expected. By mid 1916, the company's situation was favorable enough for them to purchase the Wolf group of mineral claims, located 2½ miles up-river from the Dolly Varden claims. The vendor was Donald Cameron, a prospector, and the consideration was $50,000.00. Four additional claims on the Alice Arm waterfront were purchased for a wharf, ore bunkers, warehouses and other facilities at tidewater.

The main workings at the Dolly Varden Mine were at the 1700-foot elevation, about 700 feet above the water level of the river. Using small diamond drills that had been carried in on the miners' backs, McGinnis and his crew made outstanding progress in developing the mine. By September of 1916, work completed that year

Ole Evindsen photographed these workers in 1915 as they rested during assembly of the first diamond drill at the Dolly Varden Mine. Left to right, Charlie Swanson, who received one-fifth interest in the mine for a grubstake, William McLean, of the Taylor Engineering Company, Gus Pearson, brother of Ole Pearson, one of the original four discoverers of the mine, and a Mr. Jacobsen, a miner.

consisted of 1200 feet of tunnels and raises and 5500 feet of diamond drilling which was about double the work accomplished the previous year.

The ores of the Dolly Varden Mine were beautiful to behold. They were described as "blue-tinged and white quartz with pyrite, galena and native silver, argentite, zinc-blende and rhodonite." (2.1) Even one who has not known the excitement that comes from the discovery of vast mineral wealth can understand the feeling of optimism that must have pervaded the activities of everyone connected with the Dolly Varden Mine from Joseph Hubbard and the stockholders to the lowest-ranking laborer. In the summer of 1916, the future of the Dolly Varden Mine seemed assured but transportation had to be provided before full-scale mining could begin.

THE TAYLOR ENGINEERING COMPANY

Left:
Ole Evindsen got this shot of miners Axel Norman and Jack Marshall striking the first high-grade ore at the Dolly Varden Mine in 1915.

In 1916 the Taylor Engineering Company was a Vancouver, B.C. firm of consulting engineers and equipment dealers engaged in construction for the Granby Consolidated Mining, Smelting and Power Company's huge mine and smelter complex at Anyox, about 18 miles by water from Alice Arm. Anyox, in the native peoples' "Hidden Creek" language, was to become the largest copper smelter in the British Empire. The Taylor Company had previously installed some smelter equipment and in 1916 their company was involved in constructing a hydroelectric plant. The company was also under contract to install a Marcus Tipple or pithead plant at the Granby Company's coal mine at Cassidy, on Vancouver Island. This coal mine would provide coking coal for the blast furnaces at Anyox. The president and managing director of the Taylor Engineering Company was Alfred J. T. Taylor of Vancouver.

Alfred Taylor was a young mechanical engineer who had come up in his profession through ambition and sheer hard work. He was born in Victoria, B.C. in 1887, the son of the Reverend Dr. D. W. Taylor. In the son's career one can see much of the austere, conservative Protestantism that must have characterized his father. Certainly, if hard work led to the path of salvation, Alfred would have been assured of a place in the promised land.

Alfred left school at 14 to become a machinist's apprentice. Four years later he became a mechanic with the Canadian General Electric Company. After a year with that firm he served as a works engineer with the Canadian Explosives Company. During this early period he studied engineering at night by himself or with a tutor. Later, he moved to Vancouver and worked for a year with a consulting engineer before moving to the Schaake Machine Works in New Westminster as assistant manager. He then became an assistant to H.W. Petrie, a Toronto equipment dealer.

He traveled to England for 6 months to continue his studies and to become familiar with British engineering practice. In 1912 he founded Taylor and Young, Consulting Engineers, which he continued for 2 years. Alfred Taylor was a member of both the American Society of Mechanical Engineers and the American Society of Electrical Engineers. After completing his technical education, he formed the Taylor Engineering Company in Vancouver in November of 1914 with his brother E.G. Taylor as his business partner.

The Dolly Varden Mines Company first became acquainted with the Taylor Engineering Company when R.B. McGinnis purchased equipment for the mine from them in 1915. The presence of the Taylor Company at Anyox made it an ideal choice for McGinnis. He had seen the Taylor work at the Granby Company's complex and had come to know William McLean, the Taylor Company's resident engineer in charge of construction. The firm was also known to C.C. Lindberg, a consulting engineer employed by the Dolly Varden Company to advise them on the best way to proceed with the development of their properties.

TAYLOR AND HUBBARD: 1916

Taylor and Hubbard finally met when Joseph Hubbard came to Victoria in March of 1916 in connection with arrangements for the Dolly Varden Company's incorporation as an extra-provincial company. Whether by coincidence or design, Alfred Taylor also happened to be staying at the Empress Hotel where Hubbard was quartered. One morning Taylor telephoned the president of the Dolly Varden Mines Company and the two had breakfast together.

Hubbard outlined the Dolly Varden Company's progress and mentioned the difficult country in the Alice Arm area. His company was hoping to avoid the expense of building a railroad and several alternatives were being investigated. The first alternative was the pack-trail – already there. The management thought about using Caterpillar tractors on this trail but the tractors would only be able to carry limited tonnage and could not operate in the deep snow. Even carrying concentrates by air was considered but quickly rejected. Hubbard and Taylor's discussion was of a general nature but each had the opportunity to sound the other out. They did not discuss any specific proposals but did agree to meet at a later date.

They next met in April of 1916 when Hubbard invited Taylor to meet with him at the Dolly Varden Company's office in Chicago. At this meeting Hubbard explained that after looking at the various methods of transport, the Dolly Varden Company had decided that a narrow gauge railway would be the least costly method of bringing the ore down from the mine to the facilities they would build at Alice Arm for the transfer of ore to waiting ships or barges. After further discussion it was decided that Taylor would have William McLean, his resident engineer at Anyox, go over to Alice Arm and meet with Richard McGinnis. The two engineers planned to walk up to the Dolly Varden Mine over the route of the proposed railway. McLean would then submit a report to Taylor that would be used as the basis for an estimate of the cost of a railway.

THE RAILWAY RECONNAISSANCE

Left:
Alfred J. T. Taylor, founder of the Taylor Engineering Company which contracted to build the Dolly Varden Railway for the Dolly Varden Mines Company. Taylor was a hard working young man, only able to see the best in those with whom he worked. His faith in Joseph Hubbard, President of the Dolly Varden Mines Company, led to the Taylor Company's bankruptcy.

William McLean was a man of considerable engineering experience. Before joining the Taylor Company he had worked for a number of other firms and had at least 6 years experience in railway construction and survey work. When he received Taylor's instructions to proceed to Alice Arm and to conduct a reconnaissance for a railway, he was not surprised. He had met McGinnis earlier and had made at least one trip up to the Dolly Varden Mine. On this second trip he would be more concerned with the route up the valley than with the mine.

Taylor's instructions had been explicit. McLean was to proceed to Alice Arm at his earliest convenience and make contact with McGinnis. He was to examine the route up the valley with particular reference to the conditions that would be encountered in building and operating a railway. He was to obtain enough information to prepare an estimate of construction costs. He was also to consider the railway's gauge – the distance between the rails. Should it be 24-inch, 36-inch, or standard gauge? He also had to estimate how long construction would take and consult with McGinnis when he reached Alice Arm for the details. Shortly after receiving Taylor's instructions, McLean boarded the local motor launch and headed out to meet with McGinnis.

As McGinnis and McLean walked up the Kitsault Valley for the second time in a month, McLean was again impressed with the difficult terrain. For the first three-quarters of a mile the line would run upgrade from the Dolly Varden wharf to Alice Arm along a rock-cut blasted into the side of a hill using several small trestles. The line would follow the pack-trail closely for the next 6 miles until it came to Mud Creek Canyon. This would be the easiest stretch of the railway. At the mouth of Mud Creek Canyon the pack-trail climbed up steeply to the rim of the gorge on the original prospector's trail. The Mud Creek Gorge was "impassable" but it was necessary to build the railway through it. Men would have to be lowered from the cliff edge to carve out footholds so that drilling and blasting could begin. Skirting the canyon along its western rim, McLean and McGinnis dropped down to water level again at the upstream end of the defile.

As they looked back down the canyon they saw what appeared to be a green, grass-covered hillside that would be easy to grade through gravel or clay. It was, in fact, a steep rock slope covered with moss and the cut would eventually cost three times as much to complete as originally planned.

From the Mud Creek Canyon the pack-trail and later the railway followed the west bank of the river at water level for 5 miles. The main problem on this section of the line was whether to follow the numerous curves of the river or to build a shorter but straighter line of railway. If the line followed the curves of the river it would be

The lower Kitsault River Valley during low water, looking much as it must have when the Nishga Indians made forays into the area to hunt mountain goats. Ole Evindsen had already traversed this valley many times when he took this photo, but terrain such as this proved a formidable obstacle to building the pack-trail from Alice Arm to the Dolly Varden Mine and later to the construction of the railway.

A view from Deadman's Point into the Kitsault River Valley. The river is over 200 feet straight down from where the photographer stood. In order to traverse this canyon the railway had to climb to Deadman's Point from the river level, requiring 2 miles of 5½ percent grade.

cheaper to construct than if each shoulder of rock was blasted through to straighten the grade. On the other hand the cheaper, curving grade would be more expensive to maintain because of its greater length and its vulnerability to washouts at high water.

As McLean and McGinnis proceeded they came upon gangs of men working on the pack-trail. There were some long stretches only passable on foot. It was July and no work had been done beyond the 14-mile point where the upper canyon of the Kitsault began. At this stage in the journey it was necessary to climb up to the high west wall of the canyon to a place called Deadman's Point. Two miles of 5½ percent grade were required to climb from water level at mile 14 to this location. From Deadman's Point the line would run along the rim of the gorge, 150 to 200 feet above the water. From this point on to the Dolly Varden Mine it was impossible to see where the line would run. It was almost as if a formality was being attended to before the next stage of the Dolly Varden project could begin. McLean did not like the idea of trying to make an estimate on the basis of such poor, inadequate information and he said so.

McGinnis explained that the Dolly Varden Company wanted to build a railway that could bring down 20 to 25 tons of concentrated ore per day for trans-shipment to smelters down the coast. The line was to be built to maximum grade and curvature and at minimal cost. In most places it would virtually follow on the grade of the pack-trail.

McLean pointed out that the two walks up the valley had not provided enough information to permit him to prepare an accurate estimate. Such an estimate could not be made without a detailed railway survey requiring cross-sectioning to determine the shape of the walls of the canyon; running locations to determine the most feasible route; and gathering other information to establish the amount of earth and rock that would have to be removed. He stressed the difficulty in trying to prepare an estimate of the section of line above Deadman's Point without actually having seen it. A detailed railway survey of the type required to prepare an accurate estimate would take from 4 to 6 months in the field and would cost between 4 and 6 thousand dollars. McLean recommended that such a survey should be carried out and he told McGinnis that if the work could begin immediately it should be possible to finish before the end of the year.

McGinnis was unable to agree to this detailed survey. His instructions were to obtain the lowest estimate for building the line in the shortest amount of time. McLean was being asked to submit an estimate to Taylor that would be based almost entirely on guesswork.

McLean returned to Anyox feeling uncomfortable about making the estimate. He felt that building a railway with such scanty and inaccurate information on the costs and quantities of material involved was extremely foolish. Nevertheless, he was an employee and he carried out his task as directed. After further consideration and cost calculations he came up with an estimate of $113,535.00 for the 17-odd miles of railway to the Dolly Varden Mine. He cabled his figures to Taylor in Vancouver and returned to his work on the Granby Company's works.

McClean continued to feel ill at ease in the following weeks about the estimate and every time he thought more about it, he felt a bit worse. When the estimate had been made the contractor had crews at work on isolated parts of the pack-trail over a 14-mile stretch; no work had been done beyond Deadman's Point. McGinnis had mentioned that the grade beyond mile 14 would be constructed specifically for the narrow gauge railway. McLean began to wonder how well this work was being done. He was curious to know if the contractor was blasting off some of the rock corners to eliminate some of the worst curves and if the numerous swampy stretches were being filled or avoided.

McLean was also concerned about the steepness of the grade between mile 14 and Deadman's Point. Finally he resolved to go over to Alice Arm for another look at the Kitsault Trail with a view to revising his figures. This third trip took place in September of 1916. When he returned to Anyox, McLean had enough information to revise his figures upward from the original $113,525.00 to $160,222.00. He passed this information on to Taylor feeling that he was still not satisfied with his work, but that it was the best he could do under the circumstances. The figures were revised a third time before work was started because of rising materials costs due to World War I shortages.

DEVELOPMENTS AT THE DOLLY VARDEN MINE

By the end of the 1916 season in November the pack-trail had been completed from Alice Arm to the Dolly Varden Mine. The main mine camp buildings were located 700 feet above the railway grade and consisted of a number of solidly built cabins, an assay office, and several other structures. The upper levels of the Dolly Varden Hill were beginning to look like a war zone with fallen trees scattered in piles around the numerous open cuts that had been made on the surface to expose ore bodies. A 50-foot adit or tunnel was driven 500 feet from the top of the hill and a second tunnel was

Permanent buildings under construction at the Dolly Varden Mine in 1916. Ole Evindsen included his wife, Svea, in the center of this photo to give some idea of the size of the new buildings.

driven another 30 feet lower down with branching crosscuts. These tunnels were joined by an inclined, branching raise. Still lower down the hill a large working tunnel was being driven that ran into the hill about 400 feet. Good ore had been cut into at 100 feet. Diamond drill holes penetrated the area between these levels and revealed many thousands of tons of high-grade ore. There were even larger quantities of lower quality ore on the fringes of this main deposit. Good ore had also been located on the Toric mineral claim located several hundred feet below the Dolly Varden workings on the opposite side of the river. Examination revealed that the Toric ore body was probably an extension of the Dolly Varden formation at a lower level. It could be safely said that many thousands of tons of ore had been "proven-up" on the eight claims comprising the Dolly Varden properties. It was a remarkable achievement when one considers that the only equipment available until the completion of the pack-trail had been small gasoline-powered drills, hand tools, and dynamite that the miners carried in on packboards.

THE WOLF GROUP

Two miles upstream from the Dolly Varden, development work had started on the Wolf Group of mineral claims that had been purchased by the Dolly Varden Company from Donald Cameron early in 1916. These properties consisted of four claims and covered an area of 180 acres. Three ore-bearing veins had been identified. The number one vein occurred in a large vertical bluff that was about 75 feet high and from 60 to 80 feet wide. A tunnel was run into the base for 80 feet with good ore in sight over the entire distance. The ore was also traced through a number of open cuts on the surface and was known to run up the hillside for at least 500 feet. Additional veins were found 300 feet down the hillside, establishing that the ore body extended over a vertical distance of 800 feet. Diamond drilling indicated that these veins covered an even wider area than those of the Dolly Varden, but unfortunately the Wolf ore was of a lower grade. It would be most economic to concentrate the ores ten-to-one before shipment. The company's plans included constructing a concentrating mill and building a dam on nearby Trout Creek for a hydroelectric power plant to supply both the Wolf and the Dolly Varden with electricity. (This hydro plant was built in 1920.) The railway could be extended from the Dolly Varden to the Wolf on a gentle grade so that low-grade ore from both mines could be treated together.

The Dolly Varden and Wolf properties constituted a huge reserve of ore that could sustain extensive mining operations for many years. The key to their development lay in the completion of the railway to provide low-cost transportation. As 1916 drew to a close the future of the mines on the upper Kitsault looked better than ever.

THE DECISION TO BEGIN RAILWAY CONSTRUCTION

By January of 1917 the construction of the railway was assured. Although the Dolly Varden directors had originally wished to avoid the expense of a railway, it was now clear that they would have to expend considerable money to place the mines in production so that they could generate some return on the stockholders' investments. Although a railway was necessary, they were determined to build it at the absolute minimum cost. This meant constructing a 36-inch narrow gauge line built for about

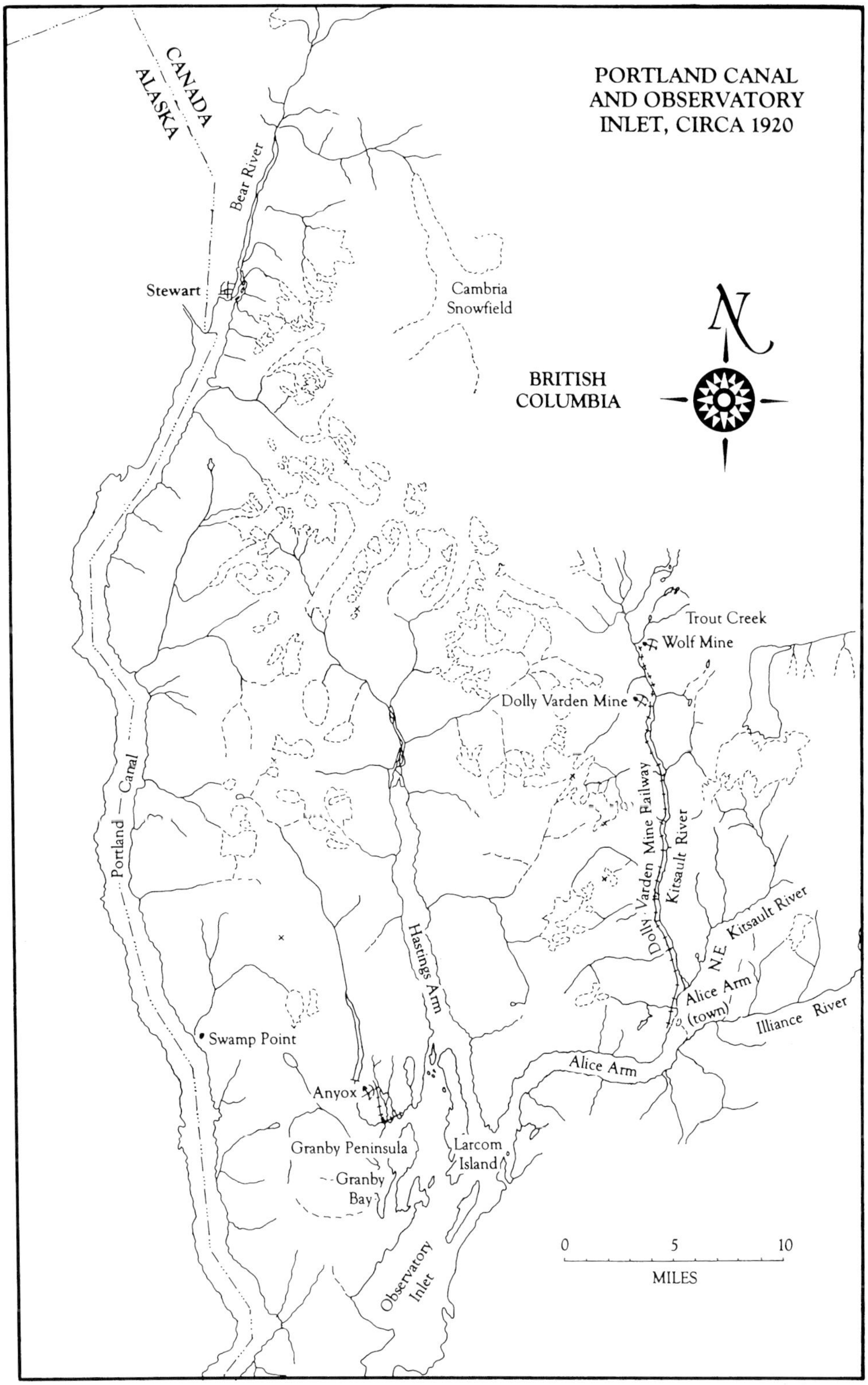

half the cost of a standard gauge line. Because of their experience and knowledge of operating conditions in the northwest coast area the Taylor Engineering Company was the logical choice to build the railway.

Before actual construction could begin, a charter for the railway had to be obtained from the Provincial Government. Also, materials and equipment had to be procured at a time when they were in short supply because of shortages created by World War I. It was imperative that Hubbard enter into a contract for constructing the line that would be based on costs acceptable to the shareholders.

Construction could not begin until the snow was off the Kitsault Trail in late April or early May. There was much to be done in the intervening months. Hubbard was determined to move with all possible speed, so he contacted Alfred Taylor in Vancouver and told him to start ordering the equipment and materials needed to build the railway. Taylor placed the first equipment order in January of 1917. It included a narrow gauge locomotive that would be used to construct the line.

At the same time, Hubbard contacted L. W. Patmore, his Vancouver solicitor, and asked him to make a formal application to the Provincial Legislature for a railway charter. Interestingly enough, this was not the first railway charter for a railway in the Alice Arm area.

On May 20, 1898, the Alice Arm Railway Company had been granted a charter to build a railway from a point at or near Alice Arm, on Observatory Inlet, thence by way of the Nass River to some point on the Stikine River at or near Telegraph Creek. A second section of the railway would then be built from Telegraph Creek by the most convenient route to the south end of Teslin Lake.

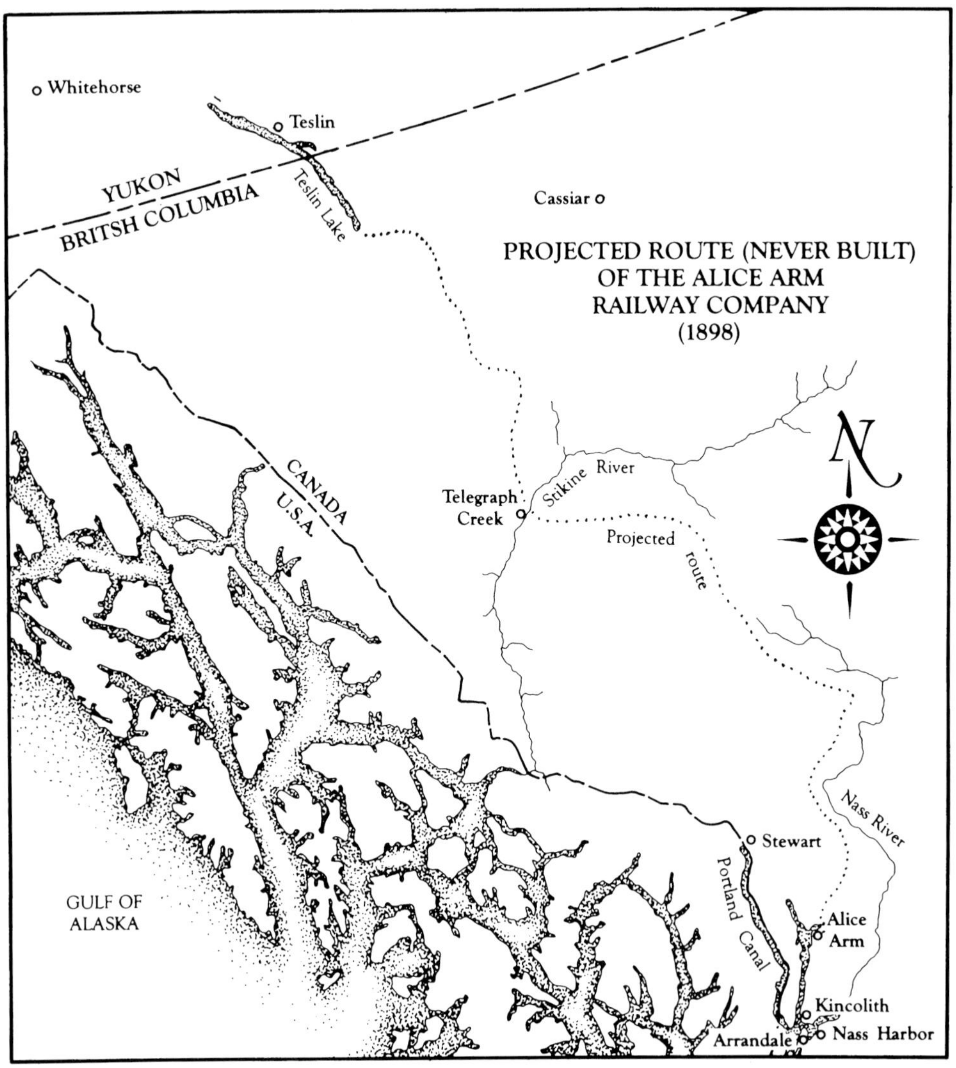

The fact that no pass was even known to exist at that time from the upper Kitsault watershed to the Nass watershed was no deterrent. It was in 1908, 10 years later, when Joe Wells and William Dilworth finally crossed over the divide between the two rivers and established the existence of a pass. The granting of charters for railways through unknown country was a widespread practice in British Columbia at the turn of the century. It was part of a game called "charter mongering" that resulted in granting charters for nearly every route a railway might conceivably wish to build. The "charter monger" then sold his charter to any legitimate railway builder who might wish to build a line through the area covered by his charter. This practice led to a number of celebrated legal cases. Finally in 1927 the Defunct Railway Companies Dissolution Act was signed into law to remove from government records the names of 128 railways that had been granted charters but had never been built. While all of those railway names were not the work of "charter mongers," it is reasonable to assume that many of them were. Fortunately the charter applied for by the Dolly Varden Mines Company was legitimate.

With the formalities of application for a charter in the hands of Patmore, Hubbard directed his efforts towards contracting for the construction of the railway. He drew up a document that he felt was appropriate and sent it to McGinnis and Patmore with instructions to present it to Alfred Taylor for his perusal and signature.

3281
OMINION
PHOTO CO.

3

The Dolly Varden Mines Railway Contract

In mid-January of 1917, Joseph Hubbard contacted Alfred Taylor through his Vancouver attorney, L.W. Patmore. Patmore conveyed Hubbard's message that it looked as if the Dolly Varden Company would be successful in obtaining a charter for the railway and that the Dolly Varden Company was anxious to start stockpiling railway supplies and materials as soon as possible. Even if the charter were not granted, the company could always resell the materials. Consequently, the Taylor Company found itself buying materials for the railway before there was even a contract to build it.

With the upper Kitsault Valley snowed in until May, R.B. McGinnis was in Vancouver to assist in the preliminary arrangements for the railway. W. McLean, of the Taylor Company, was also in Vancouver from Anyox and was contacted by Patmore and McGinnis concerning the railway contract. For discussion purposes they presented a standard, pro forma contract and a set of standard railway specifications. Patmore, McGinnis and McLean chose 12 or 14 clauses they felt might be the basis of a contract applicable to the Dolly Varden Mines Railway project. Finally, Patmore left the two engineers alone to draw up a contract that might be acceptable to Alfred Taylor.

Several days later, Patmore and McGinnis went to Taylor's office with a printed railway contract made out for a lump sum cost of $175,000.00. Taylor had said earlier that he would not sign a contract for such a sum and he refused to sign the contract. He had good reasons for this position; there was no plan for the proposed railway and no detailed survey had been made; it had not been possible for McLean to see the entire route; sources of materials were not settled and prices were rising rapidly. In addition, McLean's original estimate had already risen from $113,525.00 to $160,222.00 and finally $175,000.00. Taylor's position left the two parties at an impasse and the contract was set aside for a month.

In February of 1917, Joseph Hubbard wired Alfred Taylor and asked him to come to Chicago to see if they could negotiate a mutually agreeable contract for the construction of the railway. Taylor arrived and registered as Hubbard's guest at the posh Chicago Club.

When discussions began the next day, Taylor was determined not to accept a lump sum contract. However, Taylor, the young, rising engineer was about to do business with an older, more experienced man who knew the art of negotiating from beginning to end. Hubbard was always an affable man and his age and experience made it easy for him to inspire confidence in the younger Taylor. Their relationship remained congenial for a long time during which Taylor confided in Hubbard and took his advice.

Left:
The 30-pound rails of the Dolly Varden Mines Railway cling to the steep hillside on the 5½ percent grade between miles 14 and 16.

Hubbard explained to Taylor, in confidence, that he was having some difficulty with his principal shareholders because several contracts he'd recently arranged had gone well over the amount they initially intended to spend. It had been necessary to put up additional funds to keep these projects solvent. The shareholders were determined that this would not happen with the Dolly Varden Mines Railway and had informed Hubbard that the only contract they would accept was one for a fixed sum. He told Taylor he had already informed the shareholders the railway would cost no more than $175,000.00, the amount of Taylor's revised estimate based on McLean's revised estimate with additional provisions to cover rising costs.

While Taylor could sympathize with Hubbard's position, he reiterated that he could not possibly agree to a contract at a guaranteed limit. As far as his "guesses" went up to that time, the railway could probably be built for $175,000.00, but he was not prepared to bind his company to such an amount. The lump-sum contract presented to him by Patmore and McGinnis the month before contained clauses applicable only to a standard gauge railway, not to a narrow gauge railway such as the Dolly Varden Company wanted to build. Taylor felt these provisions were just too stringent.

What Taylor really wanted was a contract under which he would be an agent or employee of the Dolly Varden Mines Company. Hubbard could not agree to that. Taylor finally retired to his room and spent the next day drawing up a contract he felt would be mutually agreeable to both companies. When he presented it to Hubbard for comment, Hubbard proposed some minor revisions, deleted some clauses, and then the two finally had an agreement meeting the basic requirements of both firms, except the stumbling block of the fixed-sum contract.

Taylor stated his position by saying, "Mr. Hubbard, I am afraid to take a lump-sum contract. The only contract we can possibly accept is one that provides for costs plus 10 percent fees." Hubbard replied that he had already gone on record with his directors that they could depend on the construction price of $175,000.00 and that he had no authority to accept any other contract.

Taylor's contract also provided that the Taylor Engineering Company would undertake work in addition to actual construction of the railway. This included a power plant for the concentrator, construction of loading facilities and wharves, and other general engineering work. Hubbard was prepared to compromise with Taylor on these items and exclude them from the $175,000.00 contract.

However, Taylor again said, "No." So Hubbard changed his approach by saying, "Mr. Taylor, can you give me any idea of a figure that you could possibly accept?" Taylor didn't think he could. "Mr. Taylor, could the road possibly cost more than $275,000.00?" Taylor was certain that it could not. "Very well then, I think you might execute that contract as it stands at $175,000.00 as a lump sum." Taylor was still not sure. Hubbard, growing impatient, continued, "Supposing you draw up a contract in which you guarantee to do this work for a lump sum of $175,000.00 and I will get this accepted, or will accept it, and I will give you a 'gentlemen's agreement' that if it exceeds this, I will see you are recompensed." Taylor asked Hubbard to put the 'gentlemen's agreement' in writing but he declined to do so. Taylor later testified that Hubbard told him that the executor of Hubbard's estate would be told of the agreement in case he should die. Taylor had absolute faith in Hubbard's word which finally convinced Taylor to proceed.

Taylor and Hubbard finally entered into an agreement that depended on Hubbard's word and Joseph Hubbard at no time ever denied this commitment to Taylor. The "gentlemen's agreement" was acknowledged by both men and became common knowledge among their respective employees as they later worked together on the construction of the railway.

The agreement between the two men was reached on February 23, 1917, but Hubbard did not sign the written part of it until the Dolly Varden Mines Railway had received its charter from the government of British Columbia some weeks later.

PROVISIONS OF THE RAILWAY CONTRACT

The contract for the construction of the Dolly Varden Mines Railway is a rather unimpressive four page document. Its main provisions are as follows:

February 23, 1917

The Dolly Varden Mines, Limited
403 Railway Exchange Building
Chicago, Ill.

Dear Sirs:

We hereby propose to act as your Engineers and Constructors, upon written authorization from time to time to do all things necessary to plan, design, construct and equip the following engineering undertakings in connection with your interests in the Alice Arm area of British Columbia:

Item-1- A 36" gauge Industrial Railway between your wharf at Alice Arm and the Wolf Mine, a distance of eighteen miles... We anticipate building for you under this item whatever water stations, roundhouses and sidings you may require initially for the satisfactory operation of your equipment on the railway.

Item-2- The necessary rolling stock and equipment to place this railroad in position to satisfactorily handle your materials and to your several properties, and your concentrates from your mill to your wharf, and we suggest that the following equipment in this connection is desirable:

(1) One 10" x 16" Steam Locomotive fitted for oil fuel and equipped with air brakes.
(4) Four 10 Ton cars of the Rogers Ballast type fitted with air brakes.

Item-3- Under this item you will require at your wharf a 2500 bbl. steel fuel oil tank complete with receiving and delivery lines. You will also require a special oil tank car for delivering oil from your wharf tank to a tank to be located at your mill power plant.

Item-4- A complete steam power plant to the requirements of the B.C. Bureau of Inspection, and adequate for the operation of a 100 Ton Concentrating Mill as now designed for your service by the General Engineering Company, and containing 250 H.P. in electric motors...

We will do all things necessary to plan, design, furnish and install and equip this work and all the various parts thereof for the full cost and expense incurred by reason thereof, to be paid to us in full, plus ten (10%) thereon, as our fee.

It is understood and agreed that we, in order to protect you against an excessive cost, guarantee that the total cost of the railway in accordance with the attached specifications will not exceed the gross sum of One Hundred and Seventy-Five Thousand ($175,000.00) Dollars, which sum does not, however, include water stations, roundhouses, sidings, and assumes that the railway as constructed follows the center line of the existing horse trail.

The total cost of the railway under Item-1, as set forth above, is based on the quantities of materials as shown on the attached estimate sheet. No accurate surveys of the railway are available, but an allowance of 25% in its estimated quantities has been made which is deemed sufficient.

You are to deliver free of charge to us from the wharf at Alice Arm to the several points of erection, all materials and equipment required for the Fuel Oil System and the Steam Power Plant.

As to the time of completion of this work, we confidently expect to be able to complete all work as itemized in this letter, on or before November 1st of the present year, unless unavoidably delayed by non-delivery of materials, strikes or Acts of God.

You will appreciate that until we actually commence work, we cannot complete detailed specifications on any item except perhaps the railroad, but if you should employ us under our proposal we will at once prepare and submit for your approval detailed specifications under each item, before actual construction work is commenced...

We will prosecute all work in co-operation with you and under the direction of one of your authorized representatives, of whose appointment for this purpose we shall have written notice...

A competent foreman shall be kept on the ground by us during all working hours and shall receive and comply with the instructions of your representative when necessary, and should such foreman be deemed by your representative incompetent, he shall, on the request of your representative be discharged, and we shall at once appoint another in his stead...

We agree to submit to your head office at Chicago, on or before the 5th of every month, an estimate of the amount or amounts that may be due us on the 15th of the said current month...

In the case of any dispute or failure to agree, the matter in dispute shall be limited to arbitration; you and ourselves to choose a representative and the two representatives so chosen to choose a third, and the decision reached by the three in arbitration, to be binding on all parties.

This proposal is for immediate acceptance by you in writing and when so accepted will bind our respective companies.

Respectfully submitted,

THE TAYLOR ENGINEERING COMPANY, Limited

By "A. J. T. Taylor"
President and Managing Director

ACCEPTED FEBRUARY, 1917
THE DOLLY VARDEN MINES COMPANY
"J. D. HUBBARD"
President

This document is delivered and executed in duplicate.

3308
DOMINION

4

Railway and Mining Developments – 1917

THE RAILWAY CHARTER

In the fall of 1915, Joseph Hubbard came to Victoria and lobbied the Provincial Government to build a road up the Kitsault Valley to provide access to the numerous mining properties in the area. The government refused but made a commitment to the Dolly Varden Company to denote the Kitsault Trail as a public highway and maintain it if the Dolly Varden people constructed the road at their own expense. While that was less than Hubbard had hoped for, it at least allowed the company to condemn a right-of-way without having to negotiate with the owners of any private land they might cross along the way.

There already was a trail running up the west bank of the river for 7½ miles from Alice Arm. The Dolly Varden Company surveyed a route from that point and the government listed it as a road in January of 1916. A year later the company sought permission to lay rails over this public road.

The Dolly Varden Mine needed a concentrator to reduce their ore; such a facility could not be constructed without a railway to carry in the heavy equipment required to build it and to carry out the concentrated ores.

In January of 1917, L. W. Patmore, acting on behalf of the Dolly Varden Company, began negotiations to obtain a railway construction and operating permit from the government. Other mining companies and prospectors in the Alice Arm area were aware of the Dolly Varden Company's plans to build a railway. It seemed obvious to them that the projected line would have to be built on the trail that was the only passable route up the Kitsault. Laying rails on the pack-trail would make it impassable for pack horses and access to other mining properties in the upper valley might be cut off. These other companies might also have to use the railway at the rates and conditions decided by the Dolly Varden Company. Accordingly, the citizens of Alice Arm forwarded a petition containing 64 signatures to their member of the Legislature, Mr. Frank Moberly. The petitioners asked that the railway be made a common carrier with rates controlled by the government. They wanted their rights to be considered and respected.

The Dolly Varden Mines Company had graded an 8-foot wide strip at its own expense on the 66-foot right-of-way that had also been surveyed and announced at their own expense. Now they needed government permission to lay rails over this route. Under the Dolly Varden Company's original articles of incorporation in the United States it was permitted to build and operate tramways but it was expressly prohibited from becoming a common carrier. (Nor indeed did the company want to.) In his initial letter to the Honorable John Oliver, premier and minister of railways, L. W. Patmore, the Dolly Varden solicitor, asked for a lease of only 30 years over a 25-foot portion of the published right-of-way. The company's attitude with respect to the

Left:
Four Mile Falls at Canyon Creek on the Dolly Varden Mines Railway.

petitioners and the government was that "we have a charter that enables us to attend to our business but will not allow us to become common carriers. If the people of British Columbia wish us to become common carriers, they should put through a government measure in the coming session of the Legislature authorizing us to do this and giving us a right-of-way for this purpose over a portion of the public highway." (4.1)

While the Dolly Varden Company was willing to become a common carrier of freight, it did not wish to become a common carrier of passengers. The company also wanted tax exemptions on certain equipment they would need for the railway. Although the Dolly Varden Mines Company was prepared to have its articles of incorporation in Delaware amended to include the powers necessary to become a common carrier, Patmore pointed out that the most economical alternative for them would be a simple lease of a portion of the 66-foot right-of-way as mentioned above.

In his correspondence with Premier Oliver, Patmore went to some length to point out that the horse trail/public road up the Kitsault Valley was not the only feasible route. There were other routes on the east side of the Kitsault River as well as higher up on the hillside above the trail on the west bank. The grade followed by the pack-trail, which would ultimately be used by the railway, was sited to obtain the most favorable downhill grade from the mine ore bunkers to the wharf at Alice Arm. Finally, in view of the short May-to-November operating season, it was imperative that the government give its approval for the railway as soon as possible so that construction might begin as soon as the snow was off the ground.

Patmore's negotiations with Premier Oliver were successful and, by the beginning of April, 1917, several drafts of an act granting a charter to the Dolly Varden Mines Railway had been prepared. A charter under which the company became a common carrier of freight was agreed to with the proviso that the minister of railways, at some future date, could order that passengers be carried. The government agreed to a 20-year term on the lease of a portion of the right-of-way of the public road instead of the 30-year period the company had hoped for.

RAILWAY CONSTRUCTION BEGINS

By April of 1917 the Taylor Engineering Company had procured sufficient supplies and equipment to begin constructing the railway. By April 15 the act granting a charter to the railway had been given a final reading and required only Royal Assent, the final step in legislative approval when the Lieutenant Governor, as the King's representative, signs an act into law. With the charter assured, the staff of the Taylor Engineering Company headed north from Vancouver to Alice Arm on the coastal steamer.

Assistant Engineer John Seymour arrived in Alice Arm on April 19 and was followed two days later by Chief Engineer William McLean and a large force of construction workers. Work on the railway began immediately. A small Porter 0-4-0T and several dump cars already were at Alice Arm to begin work, and by mid-May the diminutive saddle-tank 0-4-0 was busily puffing up and down a short stretch of track with loads of rail, ties and timbers in tow. The picks and shovels of the graders had the earth fairly flying along the grade. Construction was finally underway.

The railway followed the original pack-trail that had extremely sharp curves up to 39 degrees – a radius of 147 feet. After the rock-cut and fill work from the wharf up to the town was completed, the grading became easier during the next few miles through the relatively flat valley bottom. Here the workers cleared out the sides of the pack-trail, filled in the low spots, made an occasional cut through low gravel ridges, and laid the ties and light 35- and 40-pound steel rails. This work progressed rapidly and gave rise to a spirit of optimism among the graders. It looked as if the railway would be completed to the Dolly Varden Mine before the end of November, 1917, as had been forecast.

Ole Evindsen photographed the first locomotive used on the Dolly Varden Mines Railway: a Porter 0-4-0 saddle tank, shown during the construction season of 1917.

ESTIMATE SHOWING QUANTITIES AND DETAILED COSTS FOR THE DOLLY VARDEN RAILWAY FEBRUARY 1917

Solid Rock	12,500	c. yds	at	$1.15	$14,375.00
Loose Rock	9,000	c. yds	at	.65	5,850.00
Dirt	20,000	c. yds	at	.40	8,000.00
Ballasting	26,600	c. yds	at	.50	13,300.00
Grubbing	10	acres	at	250.00	2,500.00
Track Laying	20.2	miles	at	350.00	7,070.00
Bridges, Trestles and Culverts					14,727.00
Rails	660	Tons	at	50.66	33,435.60
	500	Tons	at	42.50	21,250.00
Spline Bars	23.6	Tons	for	2208.00	
	9.0	Tons	at .40	664.00	
	10.0	Tons	at .47½	1447.60	4,319.00
Track Bolts	5	Tons	at	171.36	856.80
	2	Tons	at	160.16	320.32
Track Spikes	48	Tons	at	106.00	5,088.00
Ties	38,000		at 10.00 per M.		9,120.00
	2,000		at 10.00 per M.		560.00
	20,000		at .15		3,000.00
Freight, Rails, etc.	1258	Tons	at	5.00	6,290.00
Ties					6,000.00
					156,062.72
				Contingencies 10%	15,606.20
				Total estimated cost	$171,668.92

Note: These are actual figures from contract documents. The detailed cost figures for ties, spline bars, etc. indicate that some of these materials had already been purchased or contracted for. They must have been among the materials ordered by Taylor in January of 1917.

Unfortunately, completion of the line did not depend solely on the work of the graders. As the year progressed, procurement of supplies became more and more difficult. There were steel shortages and labor problems on the docks at Vancouver and Prince Rupert. For a time the flow of supplies was stopped completely. Not even the mail could get in and out of Alice Arm.

Most of the track-laying materials used on the Dolly Varden Railway were secondhand. Some of the rails came from as far away as Texas and it was to the credit of the Taylor Engineering Company that they had been able to obtain them at all. While there was an adequate supply of rails, there was a shortage of nuts, bolts, and angle bars required to fasten them together. Laying rails with inadequate fastenings would have been dangerous except that the slow speed of the Dolly Varden's small 12-ton locomotives generally prevented derailments. In spite of all these difficulties, during the summer of 1917 the little narrow gauge slowly advanced up the Kitsault Valley toward the Dolly Varden Mine.

On July 21, 1917, after only 8 weeks, William McLean left Alice Arm and was succeeded by John Seymour as chief engineer for the Taylor Engineering Company. Seymour worked closely with R. B. McGinnis and his work was subject to McGinnis's approval. While this arrangement suited McGinnis, it soon became an enduring annoyance to Seymour who could see that McGinnis's knowledge of railway construction was extremely limited.

McGinnis, as resident engineer for the Dolly Varden Mines Company, kept in touch with the Provincial Railways Department, which would have to inspect the line and equipment on completion before an operating permit could be granted. The company was supposed to have filed a general location map before construction began. This was to be followed with detailed plans and a profile showing the gradients along the line. These requirements had not been met, and A. F. Proctor, chief engineer of railways, wrote to McGinnis to request submission of these documents. L. W. Patmore, as solicitor for the Dolly Varden Company in British Columbia, finally submitted a general location map to Premier Oliver, on July 5, 1917@. Patmore indicated that a detailed survey was under way and would be submitted as soon as possible. He optimistically concluded by informing Premier Oliver that the line was expected to be completed and ready for inspection by early October.

Ole Evindsen was out on the line watching the ballasting of the Dolly Varden Mines Railway when he took this photo of unidentified workers preparing the line to carry loaded ore trains.

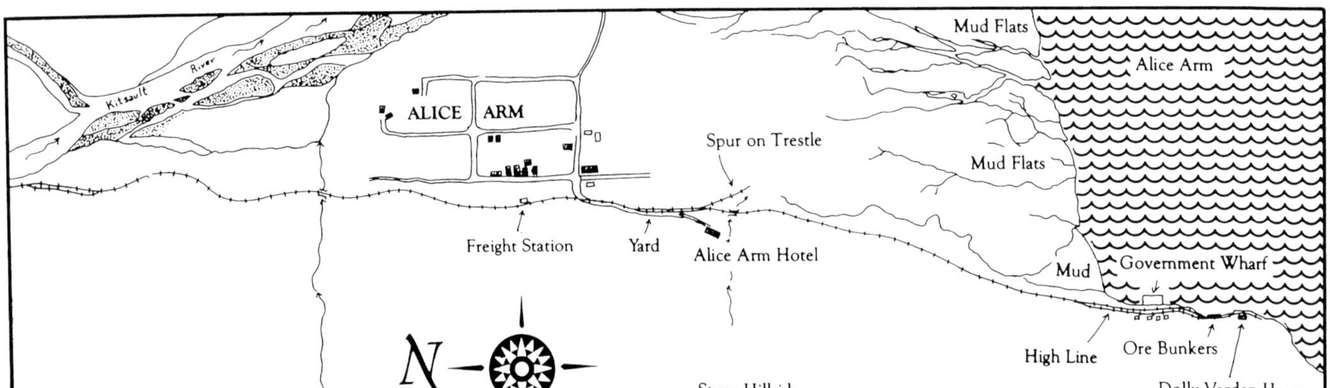

A SKETCH OF THE ALICE ARM TOWNSITE

On July 20, Chief Engineer Proctor wrote to McGinnis again with a request for detailed information on the status of railway construction. In reply, McGinnis and Seymore prepared the following chart that shows the work completed to that date:

STATUS OF RAILWAY CONSTRUCTION AS OF JULY 31, 1917

Mile	Grading	Bridges	Tracklaying	Ballast
1	100%	100%	100%	10%
2	100%	100%	100%	20%
3	100%	100%	100%	40%
4	100%	100%	100%	90%
5	100%	100%	–	–
6	100%	100%	–	–
7	60%	100%	–	–
8	70%	–	–	–
9	100%	–	–	–
10	100%	–	–	–
11	100%	–	–	–
12	60%	–	–	–
13	–	–	–	–
14	–	–	–	–
15	10%	–	–	–
16	–	–	–	–
17	–	–	–	–
18	–	–	–	–
18.5	–	–	–	–

In addition to meeting the information requirements of the government, McGinnis kept Hubbard, in Chicago, well informed on the progress of the railway. He received some assistance from C. C. Lindberg, the consulting engineer who had been retained two years earlier by the Dolly Varden Company to advise on the best method of developing their properties. Lindberg arrived in early June to examine the mine and railway and prepare a detailed report for the owners.

On examining the railway, Lindberg felt that if the rate of progress could be maintained, it would be possible to complete the line before the end of the 1917 operating season. He was not familiar, however, with the difficulties in obtaining supplies and manpower, which were not the only problems.

According to the construction contract, during the first part of each month the Taylor Engineering Company was to provide the Dolly Varden Mines Company with a list of expenditures for the Dolly Varden Railway during the previous month. The Dolly Varden Company was to pay the Taylor Company for those expenditures plus the contractor's fee of 10 percent by the fifteenth of the month. Up to September of 1917 the Dolly Varden Company had honored these conditions and had forwarded funds as required.

Unfortunately, the Taylor Company had found it impossible to hire a competent accountant. This problem, plus mail delays, had made it impossible to forward an accurate financial statement before the middle of the month as required. This meant that the Dolly Varden Mines Company had received only one financial statement prior to September of 1917.

The September statement came as a shock to the Dolly Varden people. It indicated that costs on the railway were $206,000.00 to September, which exceeded the agreed construction cost of $175,000.00 by $31,000.00. Joseph Hubbard immediately summoned Alfred Taylor to Chicago for an explanation.

Left:
The Alice Arm terminus of the Dolly Varden Mines Railway showing how the railway actually terminated near the government wharf on Alice Arm but had a freight station and yard near the town. Ole Evindsen's Alice Arm Hotel was not far from the yard. This map is based on aerial photos taken in the 1960s and includes only the main buildings, evidences of which could be seen at that time.

CHICAGO, OCTOBER 1917

On October 3, 1917, Taylor arrived in Chicago with his head office accountant and what they believed was a correct list of expenditures up to September 1. When they met with Hubbard he was critical of the way the accounts had been handled; however he reported that the Dolly Varden Company was pleased with the quality of work performed. But the inability of Taylor Engineering to provide monthly cost data was totally unacceptable to the investors he represented. Hubbard insisted that finances be kept in hand from then on and Taylor assured him that they would. Unfortunately, Taylor's assurance was unfounded, since a number of items had been billed to the Dolly Varden account of Taylor Company after the preparation of the September statement. The main reason for the delay in billing was that accounts had to be sent to Alice Arm and then back down the coast to Vancouver. Within a few days of his return to Vancouver from Chicago, Hubbard was in the equally embarrassing position of having to inform the Dolly Varden shareholders that the cost to September first was even higher than $206,000.000.

The Kitsault country did not give up its treasures easily. By the end of October the construction season at Alice Arm was drawing to a close; soon the grade would be under 15 feet or more of snow. After a general cleanup of the line, the equipment was moved down to the beach at Alice Arm to await the resumption of construction in the spring of 1918. On November 23, 1917, work ceased on the railway and McGinnis relayed the decision to close down to the chief engineer of railways in Victoria.

PROGRESS TO NOVEMBER, 1917

In spite of the problems with labor and supplies and the difficult terrain and weather, the Taylor Engineering Company was making good progress in construction of the railway. By the end of October the rails were down on the first 8 miles of the line and significant progress had been made in grading beyond that point. One hundred percent of the bridgework was completed over the first 15 miles of the line and 95 percent was finished in the sixteenth mile. This left only 550 feet of trestle to complete the bridging to the Dolly Varden at mile 17.3. The Dolly Varden Mines charter called for construction of the line to the Wolf Mine, another 2½ miles up-river from the Dolly Varden, but the Dolly Varden Mines Company had not expected to complete that portion of the line in 1917.

Development work was carried on at the Dolly Varden Mine throughout the 1917 operating season. A year earlier, in 1916, when the pack-trail to the Dolly Varden was completed, a 450-cubic foot-per-minute air compressor had been packed up to the mine and set up for operation. The compressor was gas driven and had been packed into the mine in pieces and reassembled. It had sufficient power to operate four air drills and speeded development work on the mine considerably. By the time of Lindberg's visit in early July of 1917, the #4 Tunnel (the main operating tunnel) was in a distance of 650 feet and had been connected with ore bodies on the 200 foot level above with an inclined, branching raise. A total of 1550 feet of tunneling, 350 feet of raises and winzes and 8900 feet of diamond drilling had proven the existence of at least 50,000 tons of ore averaging 44 ounces of silver per ton. At the 1916 price of 60 cents per ounce of silver, a ton of ore was worth $26.40. Thus the proven silver reserves were worth at least $1,320,000.00 in 1917. Incredibly, there were pockets of high-grade ore that contained up to 1000 ounces per ton.

While McGinnis had been a nuisance to McLean and Seymour on the railway construction, his development work at the mine was capably done. When he realized that the railway would not reach the mine by the end of the 1917 season, he tried briefly to build a makeshift road to the railhead so that some ore could be carried out to the railway. Unfortunately this work was halted by the first snowfall in late October and the Dolly Varden Mine, with her silver treasure, lay locked in the mountains along the Kitsault for at least another winter.

Left:
Ole Evindsen took this photo just below his Alice Arm Hotel in 1917. The Dolly Varden Mines Railway #1, a Porter 0-4-0 saddle tank, rounds the corner silhouetted against its own steam and smoke.

A bridge under construction along the Dolly Varden Mines Railway. The timber used was all cut locally, including the ties. The caps and stringers were hand hewn. By November of 1917 all bridges were complete on the first 15 miles of the line.

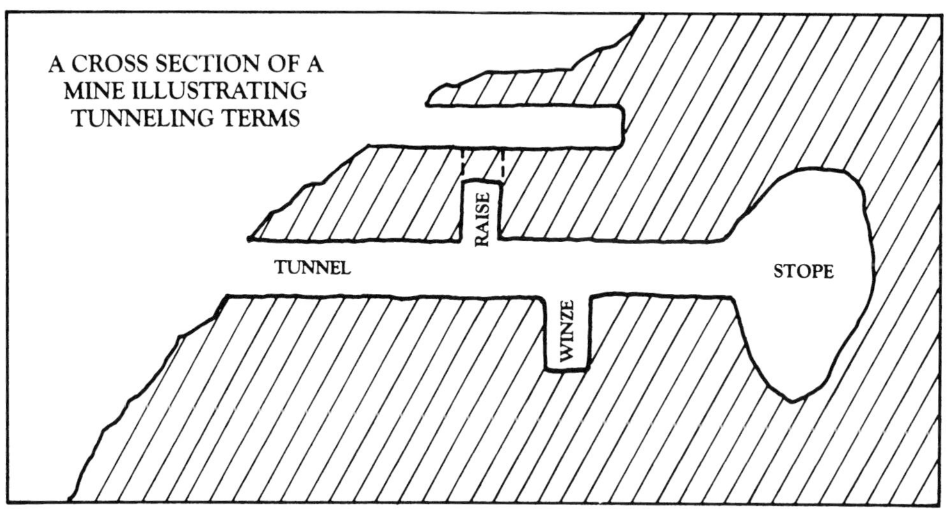

The Dolly Varden project was not one, but at least four undertakings: a mine, a power plant, a concentrator, and a railway. Considering how remote the project was, it was a highly expensive and complex series of works that were proceeding well in the face of great difficulties.

Tunneling at the mine had still only partly defined the size and extent of the ore body by November of 1917, but it was known to reach at least 150 feet below the #4

Ole Evindsen photographed a packer and two of his sturdy pack horses preparing for a trip up to the Dolly Varden Mine in 1916. In that year, along with other supplies, a 450-cubic foot-per-minute air compressor was taken apart, packed in in pieces and re-assembled. Gasoline to power this compressor that drove air-powered drills used in the mine also had to be packed in.

A photo taken in one of the stopes of the Dolly Varden Mine. The hole on the right is a "raise" to allow ore to be dropped down to a lower tunnel.

Left:
The surface buildings of the Dolly Varden Mine were at the 2075-foot elevation, about 1000 feet above Camp 8 where the railway was to terminate. The entrance to the #4 main working tunnel is just out of sight of the camera to the lower left. In the winter of 1917-1918, when work halted on the railway before its completion, development work continued at the mine, but the rich silver treasure lay locked in the mountains until the railway could be completed.

level, thereby proving the depth of the ore body at over 500 feet. Plans were well advanced in November of 1917 for construction of a hydroelectric power plant on Trout Creek, near the Wolf Mine, and were almost finished for a concentrating mill that could treat 150 tons of ore every 24 hours. The plant was designed to use a preliminary water concentration and final separation of the ore from the waste by using an oil flotation system. Preliminary tests by the General Engineering Company, the firm proposing to construct the plant, showed that up to 94 percent of the metal value in the ore could be recovered by this means.

Back in 1916 the company had figured on the pack-trail/wagon road costing between $30,000.00 and $40,000.00. Even that initial project was completed at a cost much higher than originally expected, and it was inadequate for the job. Since that expenditure, the stockholders had been induced to put up funds for the still uncompleted railway that was costing a great deal more than they were prepared to spend.

After Taylor's meeting with the Dolly Varden people in Chicago in October of 1917, the monthly advances of funds by the Dolly Varden Company against the construction contract were delayed. Acting in good faith, Taylor assumed that the money would be forthcoming and extended about $30,000.00 of his company's funds in the interim.

In November of 1917, Hubbard asked Taylor to submit a year-end statement outlining total construction expenditures. When Hubbard received word that the Taylor Engineering Company had spent at least $30,000.00 of its own, he told Taylor that he would get the money but that he would need some time to do so.

16.A

5

Early Developments of 1918

CHICAGO, JANUARY 1918

Just before Christmas of 1917, Alfred Taylor's bank contacted him to say that he had a $12,000.00 overdraft in the company account. He immediately wired Hubbard for the two to meet in Chicago in early January to seek a mutually acceptable solution to their financial difficulties. When he arrived in the "Windy City" 2 weeks later, Taylor brought with him a financial statement that indicated the Dolly Varden Mines Company owed the Taylor Engineering Company about $53,000.00. Hubbard was shocked and extremely concerned. He told Taylor he'd already advised the shareholders that they owed the Taylor Company between $30,000.00 and $35,000.00. Now he would have to approach the shareholders again for more funds and he wasn't sure whether he'd be able to get them. Taylor replied that he needed $25,000.00 immediately. Hubbard understood Taylor's predicament well enough but for the moment could not commit the funds. The meeting ended and they agreed to meet again the next day.

Early the following afternoon they met in Hubbard's office where Hubbard informed Taylor that they were to meet with several members of the Dolly Varden Company later that day. The group of speculators who were financing the Dolly Varden Company had formed a stockholders' committee at the end of 1917. Differences had arisen between those who felt they had invested enough with no return in sight and those who were prepared to spend more to see the project completed; or at the very least, the properties disposed of without loss. Naturally, those who were willing to invest more felt that they should be entitled to more profits than those who were not prepared to proceed. The group who were prepared to continue were willing to put up another $100,000.00 but Mr. George Wingfield, a prominent Nevada mining tycoon who had already loaned the Dolly Varden $150,000.00, decided to opt out. George Wingfield and Joseph Hubbard were president and vice president, respectively of the Goldfields Consolidated Mines Company of Reno, Nevada.

At the meeting, Taylor was introduced to a Mr. Hamill, the solicitor for the Dolly Varden Company; a Mr. Brewster; and an unidentified third member of the stockholders' committee. The Dolly Varden people came straight to the point — they were prepared to offer Taylor $23,000.00 cash in absolute settlement of the Dolly Varden Company's indebtedness to the Taylor Engineering Company. Unfortunately the money offered was less than the $53,000.00 that the Taylor Company had spent. In return for the $43,000.00, the stockholders' committee wanted Taylor to sign an unconditional release of the debt.

Taylor wavered. He had only been in business for 4 years and had little knowledge of the affairs now at hand. But he needed $25,000.00 immediately and he made the mistake of looking to Joseph Hubbard for advice. He and Hubbard stepped out of the

Left:
Dolly Varden Mines Railway Bridge 16A. Two company executives pose for the photographer high above Homestead Creek on this bridge constructed (like most of the railway's bridges) of local materials.

This photo shows the entrance to Mud Creek Canyon prior to the dynamiting of a rock feature the workers called "the coyote." This site is notable because a worker was killed in an explosion under this bridge. This incident happened the morning after he won a high-stakes poker game, and the money never was found near the body nor with the worker's effects, leading to a persistent rumor that the explosion was not an accident but rather murder to cover up a theft.

meeting for a moment. Taylor was desperate and Hubbard could see it. He also saw a chance to turn Taylor's respect for him to the Dolly Varden Company's advantage. Taylor looked pleadingly at Hubbard. "What shall I do? Is this the proper step for me to take?" (5.1) Hubbard assured him that it was and Taylor returned to the meeting and signed the unconditional release, losing $30,000.00 consisting of $10,000 of his company's money and another $20,000.00 in profits in the bargain.

After the meeting with the stockholders' committee, Taylor stayed in Chicago for several more days while he and Hubbard discussed the railway project and its problems in detail. In spite of previous difficulties they remained good friends. Taylor valued Hubbard's counsel and looked up to him almost as a son would to a father. The naive Taylor still felt he was being properly guided by his older "friend" who had, in fact, become his adversary. Hubbard explained to Taylor that at the formation of the Dolly Varden Company an amount not to exceed $250,000.00 was agreed upon to put the mine into production. Some development work had already been completed at the mine before the Dolly Varden Company "bonded" it from Ole Evindsen and his associates. Now after investing over $200,000.00 in the railway and more on development work, the stockholders felt they should be receiving some return. He also pointed out that the Dolly Varden Company had a further investment in the Wolf properties. The company had far exceeded their planned expenditures and the Dolly Varden had yet to ship a single ton of ore. Accordingly, the shareholders had decided to attempt to resell and recoup their investment. Under these circumstances, the Dolly Varden Company had definitely decided not to proceed with railway construction in 1918. It was a decision that disapointed both Taylor and Hubbard. Clearly, both men wanted to see the Dolly Varden Mines Railway project to its conclusion; if only a way could be found to do it.

The Taylor Engineering Company was still involved in construction work for the Granby Consolidated Mining, Smelting and Power Company at Anyox. By 1918 the Taylor Company had constructed a dam, power house, blast furnace, and a number of smaller engineering works. (See Appendix B for a description of the Granby operation). They were about to construct a coal tipple at the Granby Company's coal mine at Cassidy, on Vancouver Island. During this construction work, Taylor met frequently with F. M. Sylvester, vice president and managing director of the Granby Company, and with members of his engineering staff. Consequently, Taylor knew that the Granby people were having difficulty obtaining ores with high silica content to use as flux in the smelting at their huge facility. He was also aware that there were large low-grade ore bodies with high silica content on the fringes of the main ore bodies at both the Dolly Varden and Wolf Mines. Taylor suggested to Hubbard that perhaps the Granby Company might be interested in obtaining some of these barren ores that lay in abundance along the edges of the Dolly Varden Company's deposits. Maybe the Granby Company could be induced to complete the railway in exchange for the low-grade ores at little or no cost. Hubbard liked this idea since there was no question that the ores would be useful to the Granby people.

The next morning Taylor and Hubbard met again to discuss a proposal for the Granby Company. Hubbard brought a copy of the consultant's report on the Dolly Varden properties that had been prepared by C.C. Lindberg the previous July, and another report dealing with the geological characteristics of the mine. They sat down to talk and Taylor opened with, "It seems to me that the directors of the Granby Company might be willing to advance sufficient money to complete the railway if, in return, you would either give them the ore or sell it to them at a low price."

Hubbard replied, "You can tell Mr. Sylvester that if he will put up an amount not to exceed $150,000.00 to complete the railway, we will sell him the barren ore in the Wolf Mine. They could put in their own crew and mine it as they saw fit."

Taylor was concerned about recovering the money he had forfeited 2 days earlier. Hubbard suggested that this money might still be obtained: "In the $150,000.00 we'll specify that the Granby Company will pay $35,000.00 for the rolling stock on the railway. This will pay you back the thirty-odd thousand dollars you forfeited with us at year's end."

It remained to be seen whether the Granby Company people would like the idea of completing the railway in return for the Dolly Varden's silica-bearing ores. But Taylor and Hubbard saw the proposal as an opportunity to rid the Dolly Varden Company of the need to complete and operate the railway, a way for Taylor to recover the money owed him, and a chance to complete the railway and put the mine on a producing basis at minimal cost. After the uncertainty of the previous months, Taylor was pleased at the possibility of seeing the railway project completed. He returned to Vancouver to prepare a formal proposal for the Granby Company.

THE GRANBY PROPOSAL

On January 28, 1918, Taylor forwarded his proposal with the consultant's reports attached to F. M. Sylvester at the Vancouver offices of the Granby Consolidated Mining, Smelting and Power Company. His approach was brief and direct: "During my trip to Chicago I had several interviews with Mr. Hubbard and learned of some very large low-grade quartz bodies located at the Dolly Varden and Wolf Mines; this quartz being apparently of the nature you require for your smelting fluxes. I ventured

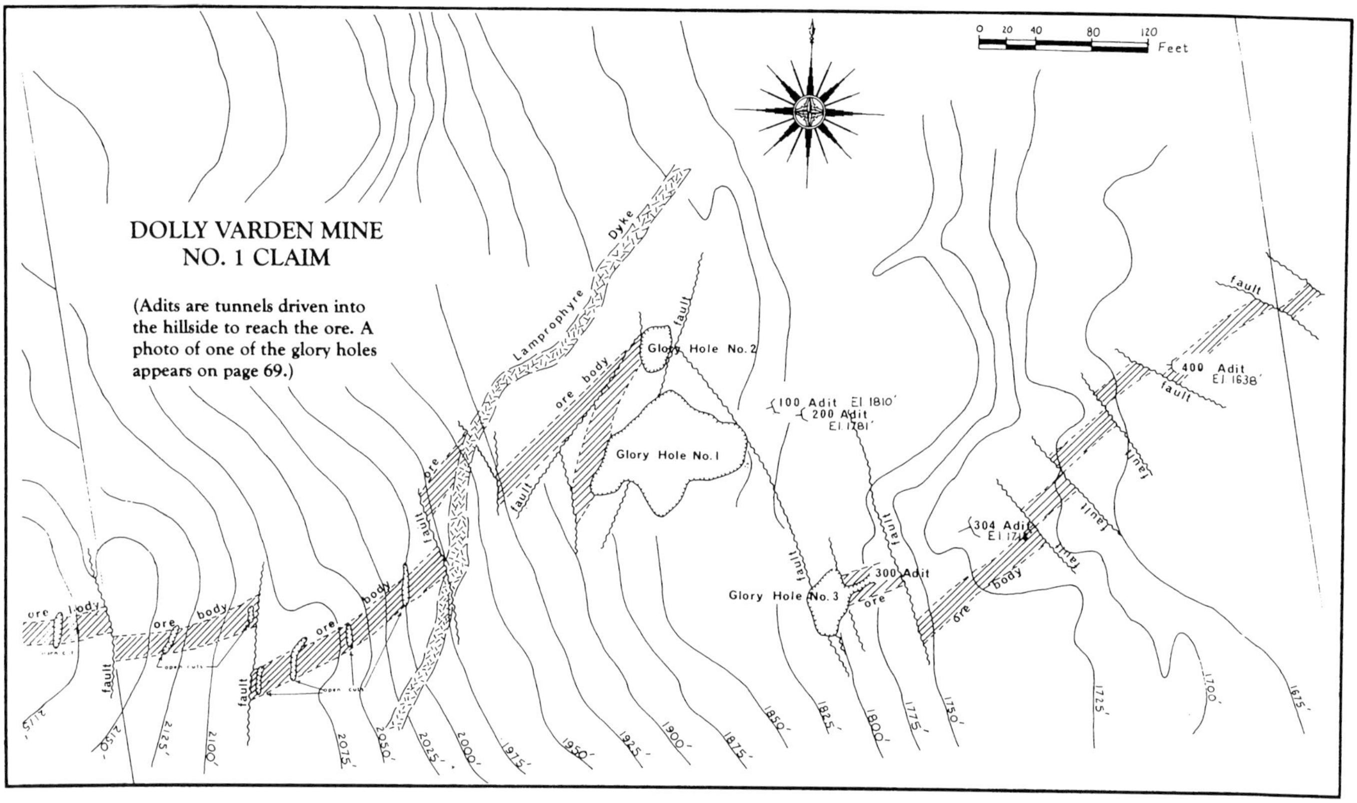

the suggestion that possibly some arrangement might be entertained whereby you and Mr. Hubbard would mutually profit, and in the discussion that ensued Mr. Hubbard kindly gave me the following facts concerning his operations in the Alice Arm District."

Wasting little time or paper, Taylor quickly pointed out that the proven low-grade ore bodies at the Dolly Varden alone were over 300 feet in depth. Every 100 vertical feet in this single low-grade deposit contained over 600,000 tons of material which, on assay, yielded marginal values of gold, 20 to 40 cents per ton; silver, 6 to 8 ounces per ton (worth only $3.60 to $4.80 per ton at 1917 prices); and traces of iron, lead, and zinc. The bulk of the 600,000 tons was silica (74.4 percent), the material required for fluxing. Continuing, Taylor described the railway and indicated that the sub-grade had yet to be completed between surveyors stations 900.40 and 912.00 (each station was 100 feet in length so this was mile 17.05 to mile 17.27). The Dolly Varden Company's investment in the railway was given as $347,000.00. The proposal to the Granby Company was for that firm to take over and complete the railway during the 1918 operating season at a cost not to exceed $150,000.00, including $35,000.00 for the rolling stock.

After completing the railway the Granby Company would retain a "substantial interest" in the railway and in the quartz ore bodies mentioned. If the project were later to prove undesirable, the Granby Company could arrange for the return of their investment and interest out of the revenues of the Dolly Varden and Wolf Mines after production was underway. Taylor even went so far as to suggest that it might be possible to run a large tunnel directly into the mountain from the railway grade at the 1000-foot elevation. This tunnel could be made large enough to accommodate the railway ore cars and would intersect the large quartz bodies at a point 500 feet in from the tunnel portal.

In conclusion, he advised Sylvester that it might be to his advantage to stop over in Chicago on his next trip east and discuss the entire situation with Hubbard and the members of the shareholders' committee.

Taylor and Hubbard were correct in assuming that Sylvester might be interested in the silica ores proposal. A month later Sylvester met with Taylor and Hubbard in

The Dolly Varden Mine Railway grade was cut out of the face of this cliff in the upper canyon of the Kitsault River Valley. This scene is on the high line above mile 16.

Left:
Ole Evindsen used his share of the money he received for his interest in the Dolly Varden Mine to build the Alice Arm Hotel. Here in 1918, Evindsen is a prosperous hotel owner. The mine was still under development and the railway was not yet finished and already in financial difficulties. In the same year the mine had yet to make a profit and construction on the railway had ceased when the Dolly Varden Mines Company refused to advance more money.

Seattle and, in March of 1918, the Granby Company took an option on the railway and on both the Dolly Varden and Wolf Mines. The option included all of the ores on both properties and, if the sale was later finalized, would include all of the Dolly Varden Mines Company's holdings in the Alice Arm area. Further developments depended upon an appraisal of the properties by the Granby Company's engineering staff. But the 1918 operating season was approaching and, if the work on the railway was to proceed, arrangements to begin construction had to be made soon.

On April 18, 1918, R. B. McGinnis traveled to Victoria to see A. F. Proctor, chief engineer of railways, in connection with the railway charter that had been granted on May 9, 1917. It called for completion of the railway not later than December 31, 1918. He explained to Proctor that while the Dolly Varden Company had hoped to complete the railway in 1917, difficulties in obtaining skilled labor and railway supplies, mail delays and strikes on the docks had made completion impossible. McGinnis wanted to know whether or not the Department of Railways would recommend that the charter be extended to a later completion date. Proctor replied that he could see no reason why an extension could not be recommended if it were necessary.

3278
DOMINION
PHOTO CO.
VANCOUVER B.C.

6

The Decision to Resume Railway Construction

VANCOUVER, MAY 1918

Alfred Taylor and Joseph Hubbard met in Vancouver during mid-May of 1918 for further discussion on the proposed sale of the Dolly Varden properties to the Granby Company and the possibility of resuming construction of the railway. Taylor and F. M. Sylvester had indicated earlier that continuing the railway construction during the 1918 season might influence the Granby Company to buy the property. Taylor and Hubbard agreed. Hubbard then suggested to Taylor that it would be a great pity if the month of May were lost to construction work. Both men knew that the Granby engineering staff would be able to complete their evaluation of the properties and forward their recommendations to the management before the end of May but that the weather would probably permit a resumption of work a month or so earlier. Taylor said that his firm had a surplus of $50,000.00 in the bank. Since he didn't expect a month's work on the railway to cost more than $35,000.00 to $40,000.00 there would be enough money left to get through May. Hubbard pointed out that the railway would be completed whether the Granby Company bought it or not. He reminded Taylor that if the Granby people did buy the Dolly Varden properties, he'd have to expect repayment of the Taylor Company's advance from them. This expectation seemed reasonable.

When Taylor mentioned this scheme to Sylvester, Sylvester was pleased and indicated that the Granby Company would pay Taylor for the work if they should later buy the railway – provided that the line was built in accordance with their construction standards. This meant that instead of constructing a "dinky" railway capable of shipping 25 tons of concentrates per day, the line would have to carry from 300 to 400 tons per day. Many sharp curves would have to be reduced and it would be necessary to use heavier rail. All of these requirements were acceptable to Taylor, and, with Sylvester's assurance that the Granby Company would pay if they bought the line, he returned to tell Hubbard who was staying at the Hotel Vancouver.

Taylor gave Hubbard the substance of his talk with Sylvester and the Dolly Varden president was pleased that construction might proceed. Taylor required the Dolly Varden Company's consent to this in writing. It was important that this letter contained a clause stating that Taylor could expect payment from only one of the two mining companies. Hubbard wrote the following letter that later became a key document in the difficulties arising between the Dolly Varden Mines Company and the Taylor Engineering Company.

Left:
The costly rock-cut at the top of the 5½ percent grade near mile 15 that was done to reduce the last remaining curve on the line to the new minimum radius of 28 degrees. Construction foreman Maloney decided on a rock-cut rather than a trestle here because a rock-cut would be maintenance-free. This cut is 40 feet high, 100 feet long and 12 feet wide.

Vancouver, B.C.
May 14, 1918

Taylor Engineering Company
VANCOUVER, B.C.

Gentlemen:-

I understand that you have proposed to Mr. Sylvester of the Granby Company that the Taylor Engineering Company should resume construction of the Dolly Varden Railroad and carry it on pending the examination of the Dolly Varden properties by the Granby Co.; that if the Granby Company should elect to exercise its option and purchase the Dolly Varden properties, the Granby Company would reimburse the Taylor Engineering Co. for the outlay and make satisfactory arrangements with you for the completion of the road. I understand that Mr. Sylvester has indicated a willingness to accept your offer.

We are not concerned with any arrangement you may make with Mr. Sylvester in regard to the matter provided the Granby does ultimately take the properties over, but in case they should not do so, it is obvious that your relations with the present owners of the property would become important and would require serious consideration.

You are already fully informed of the limitations and conditions affecting the ownership and the policy of the owners at this time and this knowledge will enable you to readily understand the position which I am compelled to take.

The owners cannot and will not assume any liability whatsoever at this time. They will recognize neither a financial nor a moral responsibility to reimburse you or to compensate you for this work in case the Granby Co. does not buy the properties. They cannot and will not permit the properties to become subject to any lien for material, labour or other charges. If the Taylor Engineering Co. undertakes this work it will be necessary to have from them an instrument releasing the present owners from any financial or moral responsibility in the matter, and expressly waiving lien rights and protecting the property against such rights, not only as to the Taylor Engineering Co., but as to subcontractors and material men.

As against this you can rely on the following facts and assurances. If the Granby Company should not exercise its option, to take over the properties, it is obvious that the present owners will either have to reorganize, complete the railway and operate the properties themselves, or dispose of them to some purchaser other than the Granby Co.; in either event, and subject to the limitations stated below, I will personally assure you that any just claims you might have for the work done and the materials furnished will be recognized and provided for either in case of reorganization or sale to others. Compensation for such claims would include not only a reimbursement, but proper compensation for engineering and contractor's profits. This assurance, however, presupposes that all work shall be done to the satisfaction of the owners, and will have to be approved by their Engineers who would be the sole judges as to whether and how far the work would be acceptable.

You may safely assume that the Dolly Varden properties with the railway constitute an asset of sufficient value to justify the assumption that any just claim you might have could be and would be taken care of in the final "workout" of the proposition and in any event. A payment of $25,000.00 is due on October 1st next, on the Wolf Claims, and the Goldfields Consolidated Mines Company loaned $150,000.00 at 6% which is due January 1st, 1919, and for which the Goldfields Consolidated Company is entitled to a mortgage. This mortgage has not yet been recorded and its final form is yet to be agreed upon. I cannot personally guarantee it, but I have no doubt the payment on the Wolf Claims will be made when due; in spite of the statement, the owners will not and cannot consider any arrangement which might create a lien upon the property, in case of foreclosure of the Goldfields mortgage, and failure to redeem by the present owners, arrangements could and would be made whereby you could redeem from such foreclosure. (6.1)

Yours Truly,

(signed)
J.D. Hubbard

BUILDING THE RAILWAY, 1918

Meanwhile, back in March of 1918, in anticipation of continuing work on the railway, the Taylor Engineering Company took in a new partner and vice president in charge of construction engineering: Jasper S. Connell, B.S., C.E., was a graduate of the New York University Engineering School with 13 years of engineering experience. Connell had worked extensively on railway surveys and construction. Right after graduation he worked for the Thompson and Starritt Company, a major U.S. engineering firm, on projects in New York, Pennsylvania, and California. He came to the West Coast in 1912 and was engaged by the Crown Willamette Paper Company in the erection of major structures. From April of 1916 to March of 1918 he superintended construction of the Pacific Mills paper plant (a Crown Willamette

In the days before stop-action cameras, Ole Evindsen took this photo of a train rounding the curve at the exit from Mud Creek Canyon.

subsidiary) located at Ocean Falls, B.C. His work there included the construction of a 7-mile logging railway on Pacific Mills timber limits at Dean Channel. Connell joined as vice president and director of the Taylor Engineering Company with one qualifying share. His wide construction experience and his detailed knowledge of northwest coast operating conditions made him an ideal choice to manage the railway construction at Alice Arm.

Connell arrived at Alice Arm within 10 days of the decision to continue railway construction. He was met by O.B. Williams, Superintendent of Smelters for the Granby Company, and his assistant, a Mr. Todd, who had come over from Anyox with Connell and McGinnis. They all walked over the first 8 miles of the railway, and Connell and McGinnis continued the rest of the way to the Dolly Varden. By the time Simon Maloney arrived from Vancouver 2 days later, 80 men were at work on the erection of camps for the construction gangs.

Maloney went up to the mine to meet Connell and McGinnis and they walked back down to Alice Arm. With detailed information at hand, Connell prepared a report outlining the engineering plan as agreed to by all parties.

Connell and Williams were pleasantly surprised at both the quality and the quantity of work carried out the previous year, as McGinnis had led them to expect less. They could only find a few instances where they disagreed with the engineering judgement used. They agreed that an enormous amount of work had been carried out and were of the opinion that it couldn't have been done at less cost. However, they decided that in order to meet the Granby requirements it would be necessary to reduce all curves exceeding the 39-degree maximum of the previous year to 28 degrees, except for a single sharp curve at the top of the steep 5½ percent grade between miles 14 and 16. In the first mile of railway alone there were 34 curves, 16 of which were over 40 degrees. These were to be reduced to 8 curves with a maximum of 20 degrees. Degrees of curvature are based on the angle subtended by a chord 100 feet long. A 40-degree curve has a radius of 143 feet and a 20-degree curve has a radius of 286 feet. Thus a 40-degree curve is sharper than a 20-degree curve. It was going to be necessary to almost entirely reconstruct the lower portion of the railway adding greatly to the cost of the line.

When work began in 1918, rails had already been laid for 8 miles. Most of the rough grading, with the exception of the large rock-cuts, had been completed to the Dolly Varden Mine, but only the first 4 miles had been ballasted. Nearly all of the bridging was in place except for a large trestle over Homestead Creek near mile 16. While it might appear that the contractors had almost accomplished their goal of reaching the Dolly Varden, the line was actually chartered to be built as far as the Wolf Mine, another 2½ miles up river. However, the Wolf Mine could not possibly be reached in the time available in 1918 and the immediate need was to reach the Dolly Varden so that ore could be shipped. Connell concluded that he could probably complete the railway to the Dolly Varden Mines by October 1, 1918 – if all went according to plan. Unfortunately, this did not happen. In May and June of 1918, no one connected with the project had even imagined the problem that arose over the next few months.

The only source of gravel for fill and ballast was a pit located just beyond mile 4. Maloney, the construction foreman, decided to start at the pit and work back to the wharf, completing the lower end of the line first. So extensive was the realignment that only about 25 percent of the grade completed the first year could be used. Meeting the Granby Company's requirements added enormously to the overall cost of the line.

Resuming work on the railway was good news for the little town of Alice Arm. The merchants, townspeople, and owners of the many mines along the railway were glad to see the work proceeding because of the benefit the railway would bring the area. The work was undertaken enthusiastically and by June 27, 1918, Maloney wrote a letter to Taylor in Vancouver stating that "of the first 4 miles from the wharf to the ballast pit, the first, third, and fourth miles have all the rock work done and need only some train filling and throwing the track into the new line and ballasting it up." (6.2)

With the completion of work on the lower end of the line, the construction crews moved to the middle and upper sections – the most difficult work.

In the meantime, events elsewhere were beginning to affect the project. When Taylor agreed to proceed with the work in May, his company had a surplus of $50,000.00 which was only enough money to carry on for one month. By the end of June he had exhausted these funds and accumulated a debt of between $80,000.00 and $90,000.00 — 30- to 40-thousand dollars more than he had available. Taylor and Hubbard were both counting on the completion of the Granby engineers' report on the Dolly Varden properties and on the sale of the mines and railway. Early in July, Hubbard traveled to Vancouver to meet with Taylor and Sylvester. When he arrived,

The Dolly Varden Mines Railway grading crew stop to have their picture taken somewhere in the depths of the Kitsault River Canyon.

Ole Evindsen photographed the tracklayers in the distance finishing up a section of track during the Dolly Varden Railway reconstruction of 1918 in which the right-of-way was widened from 8 to 10 feet and sharp curves were eliminated.

Taylor immediately pressed for financial relief. The Taylor Company needed between $60,000.00 and $70,000.00 immediately and additional funds within a very short time if work was to continue. If this money was not forthcoming, the Taylor Company would close down work on the railway.

MONEY PROBLEMS, JULY 1918

Hubbard talked around Taylor's pleas for money and told him that the discussions with the Granby Company were going well and that there would be news within a day or two. By the end of the week, there was no news. Taylor and Connell decided to tell Hubbard that he had 2 days to provide the monies required or they would stop work on the railway.

When Taylor told Hubbard about this decision, he was put off again. "It looks like the Granby deal will be consummated in the near future. Let it go for a day or two." This time though, Taylor wasn't to be put off. He pressed Hubbard hard for the money. Hubbard told him it would take at least 10 days to get the money from the Dolly Varden people in Chicago. The best he could do in the interim was to suggest that Taylor try to borrow money from the Granby Company. Taylor approached F.M. Sylvester regarding this idea and a meeting was arranged for the morning of July 18. Present were Alfred Taylor, Joseph Hubbard, Jasper Connell, F.M. Sylvester, Vice President and Managing Director of the Granby Company, with O. B. Smith, Manager of Mines, and a Mr. Wooster, Treasurer of Granby.

When the Taylor Company had agreed to open construction for the 1918 season with its own funds, it had done so in anticipation of a decision on the proposed sale by the end of June. By mid-July the Taylor Company had committed nearly $90,000.00. Financing had been arranged for a short period only as it was anticipated that the sale of the mine and railway would be completed. Taylor told the assembled group on July 18 that he would close down the work unless he could get $60,000.00 to $70,000.00 by the next day. Sylvester pointed out that "the Granby Company is not a banking institution" and suggested that Taylor try his bank. Realizing that perhaps he was being a bit hard on Taylor, he reflected for a few moments and came up with a temporary solution to the Taylor Company's money problem.

In July of 1918, the Taylor Company was constructing a coal mine tipple for the Granby Company at Cassidy on Vancouver Island to provide coking coal for the Anyox smelters. Sylvester suggested that the Taylor Company had a number of Granby contracts and that the Granby Company might be able to prepay for certain machinery at the Cassidy Mine in advance of it being shipped. The sum made available in this fashion was only enough to keep the work going for another week or two and Taylor was obliged to sign a note placing his company even further in debt.

Sylvester pointed out that the funds would last only a week or two and asked Taylor what he would do when the money was exhausted. Before he could answer, Hubbard interjected that 2 weeks would give him time enough to get back to Chicago and, although he couldn't get the money in less than 3 weeks, he could get it within 30 days. Taylor finally said, "Well, the assets of the Taylor Engineering Company are worth that amount, so we will give you the note that Mr. Wooster has suggested."

Sylvester then asked what security the Granby Company would have on the principal and Taylor replied that he assumed they would have a lien on the railway. No one seemed sure if this was so or not. After some delay, Sylvester phoned his lawyer to inquire and the meeting was adjorned with Taylor to receive the money.

As Sylvester had suggested, the funds advanced lasted only a few weeks and, when they were gone, the Taylor Engineering Company was short of funds and once again on the verge of shutting down. In desperation Taylor wired Hubbard who was now in New York to meet with the Granby Company's directors in connection with the proposed sale:

NIGHT LETTER

J.D. HUBBARD
HOTEL BELMONT, NEW YORK

INFORMATION OBTAINED TODAY CONFIRMS NECESSITY OF APPLYING FOR TWO HUNDRED THOUSAND DOLLAR LOAN AS LATEST ESTIMATES FROM ALICE ARM INDICATE PROBABILITY OF ROAD TO VARDEN COSTING THIS AMOUNT WE REQUIRE SIXTY THOUSAND AUGUST SEVENTH IN ADDITION TO AMOUNT ALREADY ADVANCED BY GRANBY ORE CAN BE HAULED OVER ROAD BY OCTOBER FIFTEENTH PERIOD WILL WIRE CHICAGO CLOSER ESTIMATE OF COST TOMORROW BUT CANNOT BE SURE OF FIGURES UNTIL CONNELL REACHES ROAD SUNDAY

A.J. TAYLOR (6.3)

Later the same day he wired Hubbard again and indicated that he had discussed the matter with his solicitor. The solicitor advised against the idea of a loan and suggested a second mortgage taken out by the Dolly Varden Company would be the best way to raise funds. Taylor stressed that he must have funds by August 7. He also wanted instruction from Hubbard by August first as to whether he should close down the work or continue. Hubbard's reply to Taylor's plea was to inform Taylor that his meeting with the Granby people in New York had been postponed for another few days. This reply didn't help and when Taylor received it he wired again:

NIGHT LETTER

J.D. HUBBARD
UNIVERSITY CLUB
NEW YORK

UNTIL CONNELL ARRIVES ALICE ARM SATURDAY AND GOES OVER LINE IMPOSSIBLE GIVE COST TO COMPLETE PERIOD DUE GENERAL POSTAL STRIKE WHICH HAS TIED UP ALL MAIL FROM JOB FOR PAST FORTNIGHT CANNOT GIVE COST TO AUGUST FIRST AND INFORMATION AVAILABLE BY WIRE INSUFFICIENT BUT IMPERATIVE WE HAVE ASSURANCE BEFORE FRIDAY NIGHT THAT AT LEAST SIXTY THOUSAND DOLLARS AVAILABLE HERE AUGUST SEVENTH TO COVER CHEQUES ALREADY ISSUED AND FURTHER ASSURANCE THAT FUNDS FOR ALL OTHER COMMITMENTS WILL BE AVAILABLE BY AUGUST FIFTEENTH OR GUARANTEED BY GRANBY COMPANY OTHERWISE COMPELLED TO CLOSE DOWN SATURDAY MORNING SYLVESTER UNABLE TO MAKE FURTHER MOVE REGARDING LOAN UNTIL MATTER REFERRED TO HIM BY NEW YORK.

A.J. TAYLOR (6.4)

Hubbard's responses to Taylor's letters and wires were evasive and inconclusive and continued for another month with no results for Taylor. He received no further funding from either the Granby Company or the Dolly Varden Mines Company. The Taylor Engineering Company was now in so deep financially that it could not close down the operation without bringing on financial disaster. In a wire to Hubbard on August 12, 1918, McGinnis stated that "contractor's position too acute to last long and closing down bound to precipitate trouble." (6.5) A huge overdraft was accumulating in the Taylor Company's bank account, but in spite of this situation, work was continuing in the deep canyons along the Kitsault.

CONSTRUCTION PROBLEMS

The construction of the railway was mainly done by subcontractors who were paid so much "per station" (100 feet). Simon Maloney, as foreman, was responsible for letting out the station contracts and the day-to-day direction of the work. In spite of this arrangement, McGinnis insisted on having the final word and passing judgement on Maloney's every decision. Maloney considered this an unnecessary and annoying imposition and he was greatly relieved when McGinnis left Alice Arm on August 9 for about a month.

Shortly after McGinnis left in August of 1918, Sylvester came over to Alice Arm from Anyox for an inspection of the railway. Prior to his visit, the maximum width of the railway was only 8 feet. Sylvester told Maloney that if the Granby Company was to become the owner of the property it would require a grade with a minimum width in rock-cuts of 10 feet and a width in earth cuts of 12 feet. With McGinnis absent, Maloney accepted Sylvester's statement as a construction requirement, necessary to insure that the railway met the Granby Company's needs. He began widening the entire road to these widths, adding 25 percent to the 1918 construction costs of the railway.

Coincident with the widening of the grade, the main focus of operations shifted from the lower to the upper end of the line where it was necessary to complete some extensive rock-cuts. The decision had been made earlier in the year to reduce all curves to a maximum of 28 degrees, except for one curve at the top of the 5½ percent grade near mile 15 that could only be reduced at great expense. There were two ways to reduce this curve; the first was to build a high, curving trestle from the end of the grade out over the river and then back to grade several hundred feet farther on; the second was to blast out a rock-cut 40 feet high and 100 feet long with a width of 12 feet. The initial cost of these two methods was similar, but the trestle would have its footings in the river and would present continuing maintenance problems. The rock-cut, however, once completed would be maintenance-free. Maloney opted for the rock-cut and more than 6000 tons of rock were removed within a few weeks. A similiar cut farther up the line required removing over 7000 tons of rock. These major works were initiated during McGinnis's absence and, when McGinnis returned to Alice Arm in September he was shocked at the cost of the work. He immediately wrote to Hubbard in Chicago about it.

Alice Arm
September 11, 1918

... "From the top of the five percent grade, fifteen miles up, we had intended to slight the work and get the rails in as quickly and cheaply as possible, and when I left to go to Vancouver I went over the ground with Mr. Maloney and showed him just what we wanted. Just after I left, however, Mr. Sylvester and Campbell came over the road and ordered Maloney to continue the road clear to the mine with just as good a roadbed as we had below, and Maloney followed his instructions as against mine. The result is that we have a fine roadbed and will have a good road, but it will cost, at a guess, $20,000.00 to $30,000.00 more than if we had put it through. I cannot blame Maloney much, as he does not know where he stands in this work, and Taylor is trying to please all parties, and perhaps, Granby more than us. If Granby takes the property, it is alright and if we keep it, we still have a good road, but at a higher investment than was absolutely necessary at the start. We could have changed the road later." (6.6)

The statement that Maloney followed Sylvester's instructions over McGinnis's is not quite true. Maloney later testified that he made the decision on his own initiative and that Sylvester merely made a suggestion. Ten days after his return to Alice Arm McGinnis wrote to Chief Engineer Proctor of the Railway Department to report on the railway's progress and to request a Railway Department inspection of the line which was necessary before operations could commence. McGinnis knew that, in spite of the Taylor Company's desperate financial position, they could not quit. Doubly confident, McGinnis informed Proctor that he expected the Dolly Varden Mines Railway to be completed to the mine and ready for operation by October 15. In the interim, events in Vancouver, Chicago, and New York were proceeding toward formalizing the agreement to sell the Dolly Varden properties to the Granby Company.

After the July meetings with Hubbard and Sylvester, Taylor kept in close touch with both of them and expected the sale would be completed any day.

On September 23, Maloney received a letter from Hubbard instructing him to do no work on the railway beyond the company sawmill located a short distance up the line from the Dolly Varden ore bunkers (recently titled "Camp 8"). All efforts were concentrated on reaching the mine and having the line ready for inspection by the Railway Department on October 15.

3304
DOMINION
PHOTO CO.
VANCOUVER, B.C.

7

The Collapse of the Granby Deal

On September 28, 1918, Alfred Taylor wired Sylvester in New York where he was meeting with Hubbard at the Granby Company's head offices. The reply he received encouraged him to believe that the deal would be completed on or about October 5. With that news, he went down to his bank to discuss the Taylor Company's overdraft that had reached about $70,000.00. He told the bank that his overdraft would probably be close to $100,000.00 before the deal came through. He said, "I want to take care of our payrolls for, if they (Taylor's creditors) get an inkling of the condition of the job and all rush down on us at once, the overdraft will be $150,000.00 and the thing will collapse while I am away." (7.1)

The bank instructed Taylor to get to Chicago and be "right on the spot when this deal is signed, you wire here and see that Mr. Sylvester wires his local office so that funds will be forthcoming at once." Taylor took the train to Chicago and arrived there on October 24, 1918. The next day he walked into Hubbard's office and asked if the deal had been completed. "What deal?" Hubbard asked.

"The Granby deal," Taylor replied. (7.2)

Hubbard said, "No it is not." Taylor repeated his belief that an agreement had been reached in New York and that a document would be in Chicago for him to sign. Hubbard replied, "Yes, there is a document here, but it is not the agreement that we reached in New York or anything like it." (7.3) Hubbard seemed distressed and as disappointed as Taylor. It appeared that the Granby Company was using the same questionable tactics with Hubbard that Hubbard had been using with Taylor.

A day later, Sylvester arrived at the Blackstone Hotel where Taylor was staying. Taylor met him and told him that he expected Hubbard would be contacting them. Sylvester didn't think so but Taylor persisted and Taylor, Sylvester and Hubbard met the next day at the office of Mr. Hamill, the Dolly Varden Company's lawyer. Hamill, as the Dolly Varden Company's legal representative, formally told Sylvester that the Dolly Varden Company could not accept the document prepared at the New York meeting. Changes would have to be made. The four talked for some time, but no agreement was reached on the changes. They finally retired to the elegant Chicago Club for lunch. Sensing that perhaps Hubbard and Sylvester would like to be alone, Taylor excused himself.

Later that day, Taylor and Hubbard met again, and Taylor learned that Sylvester had wired a Mr. Earl, the Granby Company's chief executive in New York, outlining the changes desired by the Dolly Varden people. Earl was expected to reply by wire a few days later. It was becoming clear to Taylor that the chances of success were fading and apparently Hubbard and Hamill agreed.

Left:
A rock slide in the Kitsault Canyon nearly blocks the river. A lot of backbreaking work was required to carve out a bench for the railway along this hillside.

When Taylor met with Hubbard and Hamill the next day, he raised the issue of his company's sensitive financial position. The three men talked about the possible solutions and Hubbard suggested that if the Dolly Varden Company executed bonds to the Taylor Engineering Company and gave the Taylor Company an option on the property as the Granby Company had done, Taylor might be able to sell the property to a financial house in Chicago or New York. Such a course would, at the very least, give Taylor enough money to cover his growing overdraft.

Taylor was overwhelmed by his desperate position. "Mr. Hubbard, you are a man of large experience and I am a young man. Have you some philosophy to carry you through a crisis of this kind? It is the first one I have had. I have been three or four years in building up a reputation. I have, incidentally, made a good deal of money. I am not interested in the money. I will gladly lose all the money I have made, provided I can keep the reputation of my company." (7.4)

Hubbard gave a thoughtful reply; "Oh yes, I have had trouble like this when I was young man. These things are good for one. When you get a little distance from this, you will find it a valuable experience. At any rate, Mr. Taylor, you can absolutely rely on getting the cost of this railway."... "You have built a road to Granby specifications which is a much better road than we require for the operation of our property. The best I can suggest now is that they get together some scheme by which you can finance out of it." (7.5)

Considering their discussion the previous day, Taylor decided to ask for an option. Hubbard replied that he could not give him an option but he could deed the property to the Taylor Company with an agreement so that... "for a definite amount of tonnage, and at a price that during the life of the property will give back the cost of the railway." (7.6)

Hubbard indicated that selling bonds would probably be the best way for Taylor to finance his way out of his desperate financial position. He doubted however, that such bonds could be sold in the United States. There was a possibility that bond money might be raised in Canada. Taylor was not hopeful but he said he would give the bond idea a try.

Hubbard tried briefly to comfort him. "Now you don't want to take this thing too seriously. People have gotten into trouble like this before; the main thing is to take care of your health; if you don't sleep, go to a drugstore and get some pills or something that will make you sleep... In any case, I want to assure you that you will get your money out of this. After all, that is the main thing. Owing money a short time is not so serious as long as you finally pay it off." (7.7)

(In the history of commerce did one so gullible ever suffer so much at the hands of one so devious?)

After the kind words about Taylor's health Hubbard reminded Taylor that while a great deal could be done as long as the two of them were still talking; if matters should ever come to a fight, there was no telling what might happen.

The Dolly Varden project was to be Hubbard's last before he retired and he told Taylor that he did not want to have his last job be a discredit to him in any way. Hubbard was a seasoned campaigner in the wars of commerce and the stress over the years had exacted its toll. His health was deteriorating rapidly and at the same time he was expressing concern over Taylor's health. In September of 1918, Hubbard had less than a year to live.

Taylor was impressed with Hubbard's attitude but was uncertain how to proceed. He asked Hubbard how a bond sale might be guaranteed and was told that they would be secured against the railway and the mine. The bonds would bear interest and be retired out of the earnings of the Dolly Varden and the Wolf Mines. Taylor was still reluctant to operate the mines and the railway himself but was ready to resign himself to doing so. The meeting ended.

Taylor met with Hubbard again after receipt of the wire from Earl at the Granby head offices in New York. Taylor arrived and a short time later Hubbard came in looking quite ill. Taylor suggested that Hubbard retire and that they meet the next day but Hubbard declined, saying that he had a letter to dictate for Taylor. Hubbard paced back and forth, then retired to the outer office. The secretary typed the letter and Hubbard read it, signed and passed it to Taylor. The communication denied any liability to Taylor or to the Taylor Engineering Company.

Taylor was astonished. "This is not the understanding we have been working on."

Hubbard said, "I have written that letter and there it is for you."

"What is that?" Taylor asked.

"I have written that letter and there it is for you," Hubbard repeated. Hubbard was clearly unnerved and upset. It was quite obvious from his actions that Earl's reply to the changes in the agreement requested by the Dolly Varden Company had been unfavorable and that the Granby Company was not going to invest in the mine and railway.

Still surprised by what he had heard, Taylor put Hubbard's attitude down to his ill health. He took Hubbard's letter. "Now, I can rely on the assurances that you gave me yesterday. I can wire my people in Vancouver and we can get one of these propositions for immediate relief?"

"Yes," Hubbard replied. (7.8) The meeting ended and Taylor left for Vancouver the next day.

When Taylor arrived in Vancouver, he asked Hall Chiene (of the accounting firm Buttar & Chiene) to write to Hubbard requesting him to send a letter to the Taylor Company outlining the three alternatives that had been discussed. These were: 1) the Dolly Varden Company would execute bonds to the Taylor Company and give the Taylor Company an option on the property; 2) the Dolly Varden Company would deed the property to the Taylor Company for a return of tonnage sufficient to pay the money they had invested; 3) Taylor would sell interest-bearing bonds secured against the railway and the mine. On October 11, Hubbard replied to Chiene's request with the following wire:

NIGHT LETTER

CHICAGO, ILL
OCT. 11, 1918
HALL C. CHIENE,

BUTTAR & CHIENE, VANCOUVER, B.C.
AS CONDITION PRECEDENT TO ALL DISCUSSION IT MUST BE CLEARLY UNDERSTOOD THAT DOLLY VARDEN COMPANY CONCEDES NO LIABILITY WHATSOEVER TO CONTRACTORS OR CREDITORS AND THIS MUST BE FULLY EXPLAINED TO ALL CONCERNED SUBJECT TO THIS UNDERSTANDING AND UPON CONDITION THAT CONTRACTOR RECEIVE SUFFICIENT CREDIT TO FINANCE HIS OBLIGATIONS I AM AUTHORIZED TO OFFER THREE ALTERNATIVE PROPOSITIONS FIRST RAILROAD BONDS TO THE EXTENT OF THREE HUNDRED THOUSAND DOLLARS GUARANTEED BY DOLLY VARDEN COMPANY RETIREABLE IN FIVE YEARS WITH PROVISION FOR ANNUAL TWENTY PERCENT SINKING FUND SECOND DOLLY VARDEN COMPANY WILL DEED TO CONTRACTORS OR THEIR NOMINEES RAILROAD WITH GUARANTEE OF TONNAGE UNDER AN OPERATING AGREEMENT THIRD THE DOLLY VARDEN COMPANY WILL ASSUME THREE HUNDRED THOUSAND OBLIGATION PAYABLE ONE HALF FROM DOLLY VARDEN PRODUCTION AND ONE HALF FROM WOLF THE ONE HALF FROM DOLLY VARDEN COULD PROBABLY BE REALIZED WITHIN ONE YEAR AND FROM WOLF WITHIN ONE OR TWO YEARS THEREAFTER STOP ALL PROPOSALS ARE SUBJECT TO FIRST LIEN GOLDFIELD MORTGAGE OF HUNDRED FIFTY THOUSAND WITH INTEREST WHICH IN CASE OPERATION BEGINS SPRING NINETEEN NINETEEN SHOULD BE RETIRED WITHIN NINETY DAYS AFTER SHIPMENTS COMMENCE.

J.D. HUBBARD (7.9)

None of the three proposals offered any money to allow the Taylor Company to meet its obligations so that work could resume on the railway. Instead they made certain that the Dolly Varden Company would get its share of the moneys should the mine ever begin to ship ore.

The officials at Taylor's bank had been waiting for a wire. When it came they informed Taylor that he had a lawsuit on his hands. After going on all summer in the belief that he had a way out of his financial problems, Taylor realized there was no hope. Hubbard's wire placed the resources of the bank outside of Taylor's reach. Disaster had finally come to the Taylor Company and to Alfred Taylor personally as he had committed all of his personal property and financial resources to the completion of the railway. The bank seized all funds in the accounts of the Taylor Engineering Company in Regina, Calgary, Nanaimo, Prince Rupert, and New York. Finally, on October 22, 1918, it was all over and the Taylor Company went into bankruptcy. The work on the railway ceased with rails within 4000 feet of the ore bunkers at the Dolly Varden Mine. Over 300 workmen were out of work and destitute. Outstanding wages of $150,000.00 were due and the officials of both companies were threatened with physical violence.

THE DOLLY VARDEN RAILWAY – OCTOBER 1918

The shutdown of work came only 10 days after an inspection of the railway by R. W. Macintyre, an official of the British Columbia Department of Railways. When he arrived, construction work was in full swing except for tracklaying, which was held up for lack of rails. One hundred forty tons of rail were ordered by Maloney on September 14 but did not arrive before the work was shut down on October 22. When the rails did arrive there was no one to unload them so they were returned to Vancouver. Even if the Taylor Company had continued work after October 22, the arrival of winter would have closed down the project well short of the end of the line at the Wolf Mine, 2½ miles up the Kitsault from the Dolly Varden Mine at Camp 8. The completion of the railway would have had to wait for another season even if the money had not run out. The continuation of construction was now subject to the renewal of the Dolly Varden Mines Railway charter with an extended date for completion. As the deep snows of November and December piled in the cuts and drifted in the fills, the little trains rolled down Alice Arm from the up-river camps with the last of the construction equipment and the little railway was put to bed for the winter. The locomotives were put in the engine shed at the Dolly Varden House. The steam shovel and rolling stock were set out on the siding a short distance up the line. The steam shovel was rented and lay through the long winter unused at a cost of $40.00 per day.

After the assignment for the creditors, Taylor, acting in good faith, wired Hubbard to say that while he had lost all of his personal property, he would remain with the company until all his creditors were paid in full. He suggested that if Hubbard could make a minimum cash payment of $50,000.00 and immediately deliver bonds for the balance of the work performed with an outside limit of $400,000.00, the bonds could be held in escrow and not removed until it was certain that they could be used to remove all liens against the property.

Taylor knew that the Dolly Varden Mines Railway charter would expire on December 31, 1918 and he assured Hubbard that he would cooperate with him in seeing it renewed if the Dolly Varden Company would join the Taylor Company in making this payment and bond offer to the creditors. Hubbard later indicated that he had no authority from his shareholders' committee to enter into such an arrangement.

In October of 1918, shortly before the bankruptcy of his company, Taylor had been summoned to Victoria by the Honorable J. W. de B. Farris, Minister of Labor, for an explaination of the situation at Alice Arm. The minister's request came after the receipt of a wire from P. F. Kergin to John Oliver, Premier of British Columbia, on

The Dolly Varden Mines Railway grade approaches the Wolf Mine after crossing to the east bank of the Kitsault River above mile 18. This grade never saw a rail but was used as a tote-road.

behalf of the citizens of Alice Arm that indicated the Taylor Company was not honoring checks or issuing them to pay off their men. Many of the 300 men on the job were virtually penniless.

Mr. H. C. Chiene had been appointed receiver for the Taylor Company. Acting under Chiene's instructions on behalf of the creditors, Taylor traveled to Victoria to speak with the Minister of Labor who informed him that the workers must be paid. Considering the alternatives he decided to see if he could sell the properties on behalf of the Dolly Varden Company; and in this, he was prepared to offer his services to the Dolly Varden Company without charge. Through a friend he met Joseph Errington, a representative of the Temiskaming Mining Company of Cobalt, Ontario. The Temiskaming firm was one of the largest silver mining operators in Canada and Errington was in British Columbia looking for a silver property for his company.

THE TEMISKAMING SALE

Joseph Hubbard was wintering in Pasadena, California; Taylor and Errington decided to go down to meet with him in San Francisco in January of 1919. Before they left Vancouver Taylor had already done some preliminary negotiating with Errington on behalf of the Dolly Varden people and had reached a tentative agreement in which the Temiskaming Company would buy the mines and railway from the Dolly Varden Company for $1,100,000.00. When Taylor and Errington arrived in San Francisco they had a preliminary meeting with Hubbard. After a while, Taylor retired and Hubbard and Errington were left alone to work out the final details. Taylor was greatly relieved at the prospect of a sale at last and the recovery of his moneys. He returned to his room for the first good night's sleep he'd had in months.

The next day, Errington met with him again and told him that while a financial arrangement had been worked out, it was necessary to reduce the sale price to $900,000.00. He told Taylor that the Temiskaming Company would not deal with him directly but that he must seek repayment of the cost of the railway from Hubbard and the Dolly Varden Company. What he neglected to tell Taylor was that he and Hubbard were splitting a $50,000.00 commission on the sale. The final requirement in the arrangement was that the sale was contingent upon the renewal of the Dolly Varden Mines Railway charter with a further period granted for completion of the construction work.

Taylor was finally beginning to realize that he had been victimized by Hubbard and Hubbard's associates. He had foreseen the charter renewal stipulation by the Dolly Varden Company and was having petitions circulated in Alice Arm and elsewhere asking the government to deny the charter renewal until all outstanding wages and materials were paid for. The sum of $900,000.00 was not sufficient to pay all the creditors. The Dolly Varden Company claimed to have $640,000.00 invested in the properties. Someone had to take a loss on the transaction and Hubbard's position was that the Taylor Engineering Company should share these losses with the Dolly Varden Company. In addition, Hubbard was taking the position that his firm had repudiated all financial liability to its contractors in his letters of May and September, 1918.

When Taylor and Hubbard met later that day, Hubbard mentioned that he understood petitions against the renewal of the charter were being circulated. Taylor confirmed that they were. For the first time, Hubbard was the victim and he didn't like it. "You realize this is blackmail," he said, "these petitions have been circulating absolutely with the idea of stopping us from getting our charter." Taylor replied that it was obvious the petitions were for that purpose. Suddenly, the initiative that Hubbard had used to manipulate Taylor for nearly 2 years was gone. The petitions made certain that the Dolly Varden people couldn't trade the property until they settled the wage claims and contractor's fees. Alfred Taylor and the Taylor Engineering Company had been managed by empty assurances and half-truths from the time Taylor and Hubbard first met, but not any more.

Right:
Ole Evindsen's Alice Arm Hotel with its warm hospitality and commanding view of the harbor and town was well known along the north coast. No visit to Alice Arm was complete without a stay in its comfortable accomodations. The Alice Arm Hotel was closed early for the season in 1918 to prevent the unpaid railway workers from wintering in Alice Arm.

IMPASSE

The position of the Taylor Company and the Dolly Varden Company vis-a-vis each other and the law was quite unusual. Hubbard's letter of May 14, 1918, had denied any liability to the Taylor Company and there was doubt as to whether the Taylor Company could win a case in court. On the other hand, the charter of the Dolly Varden Mines Railway had expired without the completion of the railway and the Dolly Varden people had no recourse in the courts either. The only party to which either company could appeal was the government and, in February of 1919, they both did.

Premier John Oliver, who was also Minister of Railways, had been aware of the deteriorating situation at Alice Arm for some time and had taken some preliminary steps to protect the rights of all parties involved. He had special concern for the large number of destitute construction workers. On December 4, 1918, he wrote to A. F.

Proctor, Chief Engineer of Railways, requesting that a railway inspector be sent to Alice Arm no later than January 1 to inspect and prepare a detailed report on the status of the Dolly Varden Mines Railway project. His instructions to Proctor indicated that a discreet, low-key visit was in order. "For reasons which it is not necessary to state, it is advisable that the intention of the government to make this inspection at the date above mentioned shall not become a subject for public discussion." (7.10)

THE RAILWAY INSPECTION OF JANUARY 1919

On December 30, R. W. Macintyre, an official of the Railway Department of British Columbia, left Vancouver on the steamer, *Grand Trunk*, for Anyox where he arrived on January 2. He traveled over from Anyox on the launch, *Awake*, and proceeded to Ole Evindsen's famous Alice Arm Hotel. He found the hotel officially closed for the season but was able to stay informally with the Evindsens. When work ceased on the railway in October the hotel was closed to prevent unpaid construction workers from wintering in the little town by the Kitsault. With 300 unpaid workers laid off, the community and the Provincial Police would have had great difficulty in keeping the peace.

Macintyre, like many other visitors over the years, had relied on the hospitality and warm welcome afforded one at Evindsen's Alice Arm Hotel — and there wasn't much that went on in Alice Arm that wasn't discussed in its comfortable bar, which

SUMMARY OF COSTS IN CONNECTION WITH CONSTRUCTION OF RAILROAD AT ALICE ARM, B.C. JANUARY 31, 1919

Labor:	$ 9,810.20		
Ties			
Ballasting	97,778.14		
Bridge Work	33,079.34		
Tracklaying	12,600.00		
Solid Rock	146,131.55		
Freighting	2,631.65		
Other Material	22,193.00		
Repairing Equipment	3,461.06		
Clearing	693.00		
Erecting Camps	5,761.21		
Grubbing	6,171.80		
Sharpening Tools, Blacksmith Work and Repairing	3,319.05		
Engineer's, Time Keeper's, Walking Bosses and All Overhead At Alice Arm	14,200.00		$357,830.00
Miscellaneous:	2,050.42		
Transportation			
General Expenses	38,399.39		
Board Account	1,972.43	Cr.	
Merchandise Account	585.37		39,062.75
Equipment:	14,297.90		
Railroad Equipment			
Tool Account	6,094.32		
Camp Equipment	1,988.45		
Rails, Spikes, Track Bolts, etc.	4,563.62		
Stock Account, Powder, Etc.	3,218.60	Cr.	23,725.69
			$420,618.44
Contractor's Fee 10 percent			42,061.84
			$462,680.28

had been a gathering place for the prospectors and miners who were in from the hills until spring. Evindsen, the original discoverer of the Dolly Varden, was a careful observer of the local scene. He told Macintyre that more than $600,000.00 had been spent on the development work at the mine alone. With the investment in the railway the project was well over the one-million-dollar mark.

Macintyre had inspected the railway previously on October 12, only 10 days before the Taylor Company had assigned for the benefit of its creditors. The only changes since that time were the completion of one bridge and some ballasting and train filling. In January of 1919 most of the Dolly Varden Railway was under deep drifts of snow and it wasn't possible to carry out a comprehensive inspection of the type that could be made in good weather. In spite of this, Macintyre went up to the mine by dog sled. The only inhabitant was a caretaker who looked after the mine and kept the heavy snow off the mine buildings.

Most of the railway equipment was down at the Arm. During the previous summer, the two 12-ton, saddle-tank locomotives had been augmented by a third locomotive leased from the Granby Company. All three were locked away in the engine shed at the Dolly Varden House until spring. The rolling stock and the steam shovel were set out on the siding between the dock and the town.

Macintyre reported to his superiors that the railway grade was completed and track laid for 15 miles and "a fraction." The grading was 100 percent complete for 17 miles. It was clear that the provisions of the Dolly Varden Mines Railway Act regarding completion of the railway by December 31, 1918, had not been complied with, and this information was reported to A. F. Proctor for transmission to the Honorable John Oliver as Minister of Railways.

3298.

8

The Dolly Varden Hearing

By January of 1919, the relationship between the Dolly Varden Mines Company and the Taylor Engineering Company had reached a state of open hostility. When Inspector Macintyre returned from Alice Arm, Premier Oliver was ready to decide what steps the government might take to resolve the differences between the two parties and help the unpaid workers receive the moneys due them.

On February 3, L. W. Patmore, the Dolly Varden Company's Canadian solicitor, forwarded to Premier Oliver a draft copy of a proposed new act reinstating the charter of the Dolly Varden Mines Railway. It was imaginatively titled "The Second Dolly Varden Mines Railway Act." In his covering letter to the Premier and Minister of Mines, Patmore emphasized two points: first, the Taylor Engineering Company had refused to give a list of its creditors to the Dolly Varden Company; and second, that the draft act would make payment by the Dolly Varden Mines Company a condition and not a liability. The Dolly Varden Mines Company was not prepared to accept a charter placing any liability on them.

The proposed new act asked for an extension on the completion date of the railway to December 31, 1920 – an extension of 2 years. It further sought the appointment of a judge of the Supreme Court of British Columbia to act as a commissioner to conduct an inquiry determining:

"(a) the amount fairly or equitably payable by the Dolly Varden Company to laborers for work done on the Railway; and

(b) the grants, if any, fairly or equitably payable by the Company to persons, firms, or corporations who have heretofore furnished labor, material or supplies for the construction of said railway." (8.1)

To limit the application of these provisions, paragraph two of the proposed act stated that nothing in the act should empower the commissioner so appointed to impose any liability upon the Dolly Varden Company. Patmore's position was that the commissioner would act only to determine what the costs of constructing the railway had been. A 30-day period was to follow during which the Dolly Varden Mines Company had to decide whether to accept or reject the terms of the commissioner.

Patmore sent a copy of the draft of the Second Dolly Varden Mines Act and his covering letter to the Taylor Company. The next day Taylor sent Premier Oliver a rebuttal of Patmore's letter and a petition of his own but, before doing so, he wrote a letter to Patmore and the Dolly Varden Company refuting some of the points they had made in their letter to the Premier. Taylor accused the Dolly Varden Company, and Joseph Hubbard in particular, of inducing him to proceed with the construction

Left:
A pick and sledge hammer lie temporarily abandoned at the bottom of one of the glory holes of the Dolly Varden Mine. Daylight shines into this easy-to-mine surface pocket of ore where a grizzly permits only small pieces of ore to fall down the raise into ore cars in the tunnel below.

Ole Evindsen took this panoramic photo of his Alice Arm Hotel and the town of Alice Arm from the hillside south of the hotel. A siding of the Dolly Varden Mines Railway is just below the hotel.

work on the railway in 1918 when the financing was not secure. He answered Patmore's charge that he would not provide information in regard to the Taylor Company's creditors; "If I felt that it was merely curiosity on your part which led you to desire to know the names and addresses of my company's creditors, I should be disposed to satisfy it, but to be frank, I distrust your activities and fear that you may intend to set in motion some scheme directed to creating a faction among the creditors, advantageous perhaps to those influenced by you, but extremely disadvantageous to others and myself. And therefore I have asked for an impartial tribunal to ascertain the cost of construction, I do not see the necessity of furnishing you with the information comprised in your demand No.1."(8.2) Taylor made his position clear that he objected strenuously to the draft of the Second Dolly Varden Mines Act that sought nothing more than to make the Premier and the Legislature accomplices in depriving the Taylor Company of the compensation due for its work on the railway:

"As opposed to the principle of lawless confiscation, I would like to retrace the position I sought to establish before you on the 28th ultimo, which I respectfully submit to be the true situation in which the Dolly Varden Mines Company finds itself, and is as follows: -that the Dolly Varden Mines Company, in seeking and obtaining from the Legislature a grant of special privileges, became answerable to the Legislature for the proper and honest exercise of those privileges; that on the strength of those privileges the Dolly Varden Mines Company induced me to construct the railway which it has obtained the right to build; that I acted reasonably in assuming that the Legislature would not permit its privileges to be abused or used as an instrument of fraud or oppression; that the Dolly Varden Mines Company having so availed itself of the privileges granted to it as to lead me into expenditure of a very large sum of money for the improvement of its property cannot now be allowed to play fast and loose with the Legislature, and cast away, like an old shoe, privileges which it insistently sought and diligently utilized in the effort to secure gain for itself by the creation of railway communication to its mines." (8.3)

THE TAYLOR ENGINEERING COMPANY'S PETITION

The Taylor Company's petition asked the government to "save the Taylor Engineering Company and its creditors from being deprived of a sum upwards of $467,000.00 by the Dolly Varden Mines Company." (8.4) The letter to Taylor from Hubbard (reproduced in Chapter 6) was the most important part of Taylor's petition. In it Hubbard denied any liability on the part of the Dolly Varden Mines Company to the Taylor Company and then proceeded to induce the Taylor Company to go ahead with work on the railway.

It was Taylor's contention that throughout the period from May until October of 1918, Hubbard, in his anxiety to complete negotiations with the Granby Company, had repeatedly and continually pressed for construction on the railway. Taylor's actions had been based on these assurances, both written and verbal, that Hubbard had given him. These assurances had cost Taylor his company and all of his personal property.

Taylor informed the Premier that Hubbard, Patmore, and their associates were claiming a loss in the sale of the Dolly Varden properties to the Temiskaming Company and they were asking the Taylor Company to share in this loss (and to see that the Taylor Company would receive compensation that would minimize the Dolly Varden Company's loss). Taylor felt that if there had been no previous arrangement whereby the Taylor Company would share in the anticipated profits of the venture, then neither should they have a share in the loss.

The petition charged that Taylor had been taunted by Hubbard for his failure or inability to take the Dolly Varden Company to court in these matters. It was allegedly Hubbard's boast that he had conducted affairs in such a way as to prevent the Taylor Company from having a legally enforceable claim against the Dolly Varden Company. To Taylor's way of thinking, no man should seek profit from the fruit of another man's labors and he asked the Premier to see that natural justice was realized in all of these matters.

THE SELECT COMMITTEE

In response to the Dolly Varden Company's petition for renewal of the railway charter and the Taylor Company's counter petition seeking to have renewal denied until the contractor and his creditors had been paid for their costs, Premier Oliver placed the Dolly Varden case before a select committee of the Provincial Legislature. The select committee was composed of David Whiteside, F.A. Pauline, J. S. Cowper, John Yorston and W.A. MacKenzie. The hearing was convened on February 12, 1919, and after two exhaustive weeks the select committee handed down their recommendations to the Legislature on February 25, 1919. In addition to several hundred pages of supporting documents, the testimony ran to over 550 typewritten pages.

PROCEEDINGS

In the hearing, the Dolly Varden Company was represented by L.W. Patmore, its Canadian solicitor, and by Mr. Hamill, its Chicago attorney. The Taylor Company was represented by E.C. Mayers of Vancouver. The hearing was conducted in much the same way as a trial but many strict rules of law were waived as the committee was more concerned with finding an equitable solution for the parties than it was with imposing a penalty on one party or the other. E.C. Mayers opened for the Taylor Company as petitioners and the Dolly Varden Company's Mr. Hamill acted on behalf of the respondent.

Mayer's strategy was directed at proving in as much detail as possible just what the construction work on the railway had cost; how the Taylor Company had been induced to proceed with the work in 1918, and how well the railway had been constructed. He also sought the appointment of a board of arbitrators or a tribunal to go into the detailed costs of the work. Part of the relief sought by the Taylor Company was repayment of the amount so proved.

The Taylor Company's position was: 1) that the Dolly Varden Company had entered into negotiations with the Granby Company aimed at selling their properties at Alice Arm, and 2) that the agreement was to complete the railway in order to assist the Dolly Varden Company in selling the property or, 3) that if the Granby Company did not buy, the cost of the railway would be paid by the Dolly Varden Company or some other party who might buy the mine and railway.

Great emphasis was placed on the gentlemen's agreement between Hubbard and Taylor at the time the original contract was signed in 1917. In his cross-examination of

Ole Evindsen captured Alice Arm's Kitsault Cafe and other main street buildings on film on July 1, 1919.

Taylor, Hamill went into this agreement in some detail. Taylor and Hubbard both acknowledged the existence of the gentlemen's agreement, although each had his own interpretation of it. Hamill confirmed Taylor's understanding of the agreement as follows: "... So that your present claim is that you entered into a written agreement that the railway would be built for $175,000.00 for the purpose of inducing the directors and stockholders of the Dolly Varden Company to go ahead, while you and Mr. Hubbard had a secret agreement that the guarantee would not be binding. Is that right?" (8.5) When Taylor agreed that it was, he was accused of entering into a conspiracy with Hubbard to defraud the Dolly Varden stockholders. It would be nearer the truth to say that Taylor was a somewhat naive accessory to the fact that Hubbard was deceiving both the Taylor Engineering Company and the Dolly Varden stockholders. Taylor's reply to Hamill's accusation was that... "I don't agree that there was anything wrong in what was done. I regarded Mr. Hubbard as the Dolly Varden Mines Company, as a perfectly honest man, who would not ask me to do anything that would be in the least dishonest." (8.6) It was clear that Taylor looked at these matters in moral rather than strictly legal terms. The gentlemen's agreement was acceptable to Taylor because it was clear to him that a railway would have to be built at any cost before the mine could be brought into production.

It was also clear from Taylor's testimony that he was not aware that the stockholders' committee had assumed complete control of the Dolly Varden Company and was only keeping Hubbard on to try to sell their holdings. Taylor had always regarded Hubbard and the Dolly Varden Company as one and the same.

Continuing, Hamill discussed the widening of the railway grade by 2 feet and estimated that it had cost an additional $75,000.00 to $100,000.00. The visit of F. M. Sylvester of the Granby Company to the railway during McGinnis's absence the previous August had been costly indeed. Capitalizing on this point, Hamill tried to show that the Taylor Company had built a railway that was not called for in the agreement.

A key piece of evidence in the Taylor Company's case was the letter from Hubbard to Taylor on May 14, 1918, in which Hubbard first tried to deny any liability on the part of his company and then attempted to assure Taylor that he would have a claim in any case. It was this assurance that induced Taylor to proceed with the work on the railway in 1918.

The Granby deal failed because the Granby Consolidated Mining, Smelting and Power Company simply wanted too much from the Dolly Varden Mines Company. If their proposal had been accepted, the Granby Company would have had possession of the properties for 10 years and would then have required the Dolly Varden Company to buy the properties back for $700,000.00.

Ole Evindsen got the Dominion Day celebrants to pose for this photo on the steps of the Alice Arm "Segar Store" on July 1, 1919. In the summer of 1919, hopes were high that the railway would be completed to the mine before the end of the season in October.

The Dolly Varden Company's position was that the Taylor Company had undertaken the construction of the railway in 1917 and had failed or refused to complete it. The word "refused" was not so much used in the sense of having willfully refused but more in the sense of not having completed what they had begun.

The Dolly Varden Company took the position that the contract with the Taylor Engineering Company signed in 1917 was terminated by Taylor's signing the release at the meeting of Taylor and Hubbard in Chicago in January of 1918. Therefore, after that date, the Taylor Company was in a contractual agreement with the Granby Company. It was in connection with this release and a letter of May 14, 1918, that Mayers was able to drive home a vital point in his case. During his cross-examination of Hubbard, Mayers asked if the Dolly Varden Mines president had ever practiced law. It came out that Hubbard had conducted a law practice in Toronto for a number of years before going into the mining business. Mayers referred to the conversation he had with Taylor during January of 1918 when the two retired from the Dolly Varden stockholders' meeting to discuss whether or not Taylor should sign the release. (8.7) Mayers asked Hubbard whether or not he, as a lawyer, would have counseled a client to ever sign such a release. Hubbard was forced to admit that he had advised Taylor to do so only as a means of redeeming himself with the stockholders' committee. Mayers had maneuvered Hubbard into a position that clearly demonstrated Hubbard had used Taylor's confidence in him to manipulate the younger and less experienced man. At this point in the proceedings Hubbard was in a difficult position with his own directors as well as with the Taylor Company. Hubbard's health had suffered greatly over the months preceeding the hearing and the stress was taking its toll. Although no one knew it at the time, Hubbard had only a few months to live.

The Dolly Varden Company representatives expressed belief that in October of 1918 the railway was so near completion that it would have been finished in another few days. But, as indicated previously, the contractors had run out of rail and could not possibly have completed tracklaying to the mine before the onset of winter – even if they had wanted to. The Dolly Varden Company claimed expenditures of $270,000.00 on the railway in 1917, and about $500,000.00 in the development of the mine by the end of 1918. They now prayed for the restoration of the Dolly Varden Mines Railway charter — a boon that could only be granted by the British Columbia Legislature.

FINDINGS OF THE SELECT COMMITTEE

After two exhausting weeks of testimony, examination and cross-examination, the select committee handed down its report on February 25, 1919. The members of the committee found that, in the beginning, the Dolly Varden Mines Company had undertaken to construct a light, "dinky road" along the pack-trail. This "dinky road" would have been able to carry 25 to 30 tons of concentrates per day from the mine to tidewater. In its desire to have the railway constructed as quickly and cheaply as possible the company proceeded without any formal railway survey. If a proper survey had been carried out, the costs and work involved could have been determined much more accurately. The Taylor Company accepted a contract to build the railway on a cost plus 10 percent basis with a maximum limit of $175,000.00. This contract, however, was modified by a gentlemen's agreement between Hubbard and Taylor.

The committee also determined that difficulties with labor, the cost of materials, improved standards of construction and engineering problems led to a total cost of $275,000.00 in 1917 alone. The Dolly Varden Company honored the gentlemen's agreement during the first year of construction by paying the Taylor Company $70,000.00 of the extra $100,000.00 due for the 1917 work. They ruled that at this point there was no dispute between the parties regarding this work in 1917. The matters in dispute related only to construction work on the railway during 1918 and there was nothing that required adjudication by the select committee prior to that time. The select committee also found that the railway construction work had been carried out at a reasonable cost under the supervision of McGinnis as resident engineer for the Dolly Varden Company.

Evidence given by Macintyre of the Provincial Railway Department and Proctor, Chief Engineer of Railways, showed that the railway had been constructed only to meet minimum requirements. If it had been less well built the department would not have been prepared to inspect it upon completion for the issuance of an operating permit.

One surprise to those who had been following the Dolly Varden affair was that an organized group of creditors of the Taylor Company had made an offer of $1,100,000.00 to the Dolly Varden Company after the $900,000.00 offer made by the Temiskaming Company. Acceptance of the creditor's offer would have allowed the Dolly Varden Company to pay off all of its debts with a surplus of $50,000.00; however the company was bound by the option it had already given to the Temiskaming Company.

RECOMMENDATIONS OF THE SELECT COMMITTEE

In its award, the select committee made eight major recommendations as follows:

"1. That a renewal of the charter of the Dolly Varden Mines Railway be granted to the Dolly Varden Company upon compliance with terms and conditions set down in the draft Bill annexed thereto. (The Dolly Varden Mines Railway Amendment Act, 1919)

2. That if the Dolly Varden Company and the Taylor Company cannot agree upon the cost of construction of the Railway during 1918, such cost shall be ascertained by reference to one of the Judges of the Supreme Court of British Columbia to be named by the Minister of Railways, and the cost so ascertained, together with 10% thereof added for contractor's profit, shall be declared a debt due from the Dolly Varden Company to the Taylor Engineering Company.

3. That the sum so ascertained to be due be constituted a lien or charge upon the said Railway and upon all the assets of the Dolly Varden Company... subject only to a mortgage for the sum of $150,000.00 and interest in favor of the Consolidated Goldfields, Limited; and to the balance of purchase money due to Donald W. Cameron in respect to the Wolf Group of mineral claims.

4. That all wages due in respect of the construction of the Railway in 1918 be paid by the Dolly Varden Company within 14 days after the enactment of the legislation giving effect to your committee's report.

5. That the balance of the moneys due the Taylor Company in respect of the construction of the said Railway shall be paid within 30 days after the amount of the same has been so ascertained.

6. That if the Dolly Varden Company do not wish to assume the burden of the aforesaid payments to the Taylor Engineering Company, they must so notify the Minister of Railways within one week after the adoption of this report by the Legislature, in which event the Taylor Company shall have the right, upon the terms and conditions hereinafter *(sic)* mentioned, to acquire from the Dolly Varden Company the said Railway and all the assets enumerated in the schedule attached to the said Bill, and the Dolly Varden Company shall be entitled to receive from the Taylor Company the amount of its investment in said assets in said schedule mentioned, such amount to be ascertained, in case of disagreement, in the same manner as the cost of construction of said Railway.

7. In the event of the Taylor Company becoming entitled to acquire the said Railway and other assets of the Dolly Varden Company hereinbefore (sic) mentioned, the Taylor Engineering Company shall pay off and discharge the said mortgage to the Consolidated Goldfields, Limited, or otherwise protect the Dolly Varden Company against all claims and demands in respect of the same, and shall also pay off the balance of purchase money due to said Donald Cameron in respect of the Wolf Group of claims mentioned in said schedule, and shall also within one week after the adoption of this report produce evidence satisfactory to the Minister of Railways of the Taylor Company's ability to complete the purchase of said assets from the Dolly Varden Company.

The hillside behind Alice Arm as seen from the mud flat at the entrance to the town. The fourth and fifth buildings from the left are the buildings of the Dolly Varden Mines Railway freight station.

8. In the event of the Dolly Varden Company deciding not to pay the cost of construction of said Railway, and in the event of the Taylor Company acquiring and expecting to take over the said assets upon payment of the amount of said Dolly Varden Company's investment in the same, the latter company shall be released from all claims of the Taylor Company in respect of the construction of said railway, and shall be entitled to be paid by the Taylor Engineering Company the amount of said investment so ascertained as aforesaid upon the terms set forth in the draft Bill submitted herewith." (8.8)

These recommendations were given force of law under provisions of the Dolly Varden Mines Railway Amendment Act, 1919; which was passed unanimously and given assent on March 28, 1919. The bill provided that all charges were to be paid off in the fixed order of priority out of the proceeds from the operation of the mines, with the Taylor Company and the Dolly Varden Company being third and fourth, respectively, in the order of priority after the Consolidated Goldfields, Limited and Donald Cameron, who sold the Wolf Group of claims to the Dolly Varden Company in 1916 and who had not yet been fully paid although he had received one half of the $50,000 price.

The destitute workers who had been virtually marooned at Alice Arm when construction of the railway ceased, were first in the order of priority to be paid.

After the act was passed and given Royal Assent (8.9), a judge of the Supreme Court of British Columbia fixed the cost of construction of the railway and the 10 percent profit due to the Taylor Company at $462,628.00; and the amount due to the Dolly Varden Company for investment in the properties at $613,000.00. The Dolly Varden Company failed to take advantage of Clause 4 of the select committee's recommendations and the Dolly Varden properties were vested in the Taylor Company as provided for in Clause 6.

BACKLASH

Within a day of the select committee's report being placed before the Legislature, the Temiskaming Company and the Dolly Varden Company combined forces in an effort to have the legislature defeat the recommendations. The halls outside of the legislative chambers were filled with lobbyists bent on the defeat of the Dolly Varden Mines Railway Amendment Act, 1919.

The Temiskaming Company contended that its vital interests were involved, as they had an option to buy the Dolly Varden properties. Responding to the Temiskaming complaint, Premier Oliver informed the House that his government was not anxious to have major financial interests go away, thinking that the Province had not given all concerned parties a fair hearing. He brought in a resolution giving the select committee an opportunity to hear any further evidence. The motion passed, and the Temiskaming Company was given a day in court to present their case. It is not recorded what evidence was heard, but the committee did not change its position.

A day later the House unanimously adopted the recommendations of the select committee; passed the Dolly Varden Mines Railway Amendment Act, 1919; and defeated the combined Dolly Varden and Temiskaming forces. The unanimous adoption of the recommendations had the appearance of an ultimatum to out-of-Province investors that British Columbia firms would be protected against "aggressive and sometimes conscienceless foreign capitalists." (8.10)

The next challenge to the act came in the form of a communication from the federal government in Ottawa indicating that a petition had been received asking Ottawa to overrule the Province. The sponsor of the petition was George Wingfield of the Goldfield Consolidated Mines Company. When Wingfield realized that the select committee's recommendations protected his interests in connection with the mortgage he had earlier been given by the Dolly Varden Company, he withdrew the petition and cleared the last obstacle in the way of the Taylor Company's takeover of the mine and railway.

A view looking up the main street of Alice Arm from the mud flat adjacent to the town. The tents must have been a miserable place to live and work in the cool damp north coast weather. In the spring of 1918, over 400 workers passed through the town on their way to construction jobs on the railway. In October they came back, laid-off and unpaid, after the Taylor Company went bankrupt waiting for the promised sale of the railway to the Granby Company.

SPAGHETTI

9

The Taylor Takeover

Maj. Angus Davis, Superintendent of the Dolly Varden Mines for the Taylor Mining Company. Before becoming part of the Taylor Mining Company, Major Davis had managed a number of other mining properties in western Canada.

Left:
The Dolly Varden Mines Railway workers' cookhouse at Camp 5. The neatly set table and stacks of fresh bread signal the approach of the dinner hour.

In May of 1919, the Taylor Engineering Company and its creditors organized the Taylor Mining Company to operate the Dolly Varden Mine and Railway. The Taylor Mining Company was capitalized for $1,500,000.00. Of the $1,500,000.00 in shares offered for sale, $1,200,000.00 were sold, most to the new board of directors. Alfred Taylor became president and the new directors were: R. P. Butchard, president of the B.C. Cement Company; Hall C. Chiene, of Buttar and Chiene (trustee for the creditors); C.H. Rowlson, general manager of the Imperial Oil Company; and J. F. Tener, general manager of the Pacific Lime Company. An experienced staff was selected to complete and operate the mine and railway.

The superintendent of the mines was Maj. Angus Davis, a returned soldier who had managed the Richmond Eureka Mine at Sandon, B. C.; the Snowshoe, the #7, and the Great Sullivan Mine, all properties of the Consolidated Mining, Smelting and Power Company. Davis organized and commanded the Third Canadian Tunneling Corps during World War I. He was initially assisted by Richard McGinnis who had been conspicuous in his absence from the Dolly Varden hearing. The general business manager was Maj. D. Desmond Young, formerly adjutant of the 72nd Battalion, Canadian Expeditionary Force. Young had a wide background in business, both in Canada and the United Kingdom. His assistant was Lt. J. W. Smith. The chief mechanical engineer was Frank Sawford. Other key personnel were Tom Dalrymple, construction engineer on the railway; an assayer, locations engineer, and a storesman. Silver dreams were very much alive, and with this capable staff it appeared that steel rails would soon allow realization of the dreams.

On June 4, 1919, Alfred Taylor, as president, wrote a check for $134,500.00 to Hall Chiene, trustee for the Taylor Company's unpaid workers. The money to pay the workers was taken from the capital of the new Taylor Mining Company. The records are not clear whether the moneys owed by the Taylor Engineering Company to creditors for supplies and materials were ever paid. The fact that the workers were finally paid and that work would begin at the mine again was good news. The headlines in the Vancouver Province proclaimed "Wage Claims Paid Over" — "Taylor Mining Company Took Over Dolly Varden Mine Today" — "Mining Staff, With Locomotive Crew and Equipment, Left For North."

Going north with the staff was a cargo of rails for completion of the trackage as far as the Dolly Varden Mine bunkers at Camp 8. The mining equipment they were bringing in was most complete, and consisted of a portable sawmill, air compressor plant, rock drills, mine cars and rails. Cabins were already at the mines for the members of the mine staff. Other buildings included an assay office, cookhouse, and a boarding house for the miners. To facilitate communications, telephone connections

Maj. D. Desmond Young, General Manager of the Dolly Varden Mines Railway and the mines for the Taylor Mining Company. Major Young had commanded the Third Canadian Tunneling Corps during World War I and had been awarded the Military Cross.

Camp 8 at the upper terminus of the Dolly Varden Mines Railway. The cookhouse is on the lower right and the bunkhouses are on the left. In the distance, steam issues from the powerhouse fueled with bunker fuel. The barrels of oil were brought in on barges to the Dolly Varden House and then hauled up to the mine on the railway.

The bunkhouse at Camp 5 on thc Dolly Varden Mines Railway. The new stove in the foreground and the long woolies on the line are momentos of the cool nights experienced by the workers who lived here. The workers who stayed at Camp 5 while improving the right-of-way in 1918 were not paid until June 4, 1919. It required a select committee of the Provincial Legislature and the organization of a new company, the Taylor Mining Company, before money could be found to pay them.

Dinnertime in the Camp 8 cookhouse. The Dolly Varden Mines Railway tracks are visible through the open door.

were installed between the mine, the wharf at the Dolly Varden House and at all sidings along the railway. There was even a 42-foot motor launch, the *Gore*, which maintained communications with Anyox and Prince Rupert. Only the completion of the railway stood in the way of the Dolly Varden Mine making its first shipment of ore.

On June 23, Chief Mechanical Engineer Sawford wrote to the Department of Railways requesting that an inspector be sent to Alice Arm to inspect the railway and that permission be granted to operate the locomotives during completion of the construction work.

THE RAILWAY – JULY 1919

On July 11, 1919, William Rae, Chief Inspector of Equipment for the Department of Railways, examined the railway for the government. This was the third inspection within a year. His findings confirmed those of the select committee: namely, that the line was built to minimal standards. After the snow of the previous winter and the spring runoff, Rae was inclined to think that even "minimal standards" was an exaggeration. The roadbed had suffered severe damage from washouts in some areas and it was necessary to ballast and realign the rails. Fill and additional ballast were required in many places over the entire length of the line. Most of the secondhand rails had surface kinks or were twisted in numerous places. The ties on many of the bridges were not properly spaced. The line was equipped with knife-type switchstands not acceptable to the Chief Engineer of Railways and they would have to be replaced by proper, target-type stands.

As an additional safety measure, the Department of Railways required that all switches be packed. This procedure involved nailing planks between the rails in the switch frog and between the guardrails and running rails so that trainmen could not get their feet caught between them. Prior to that time a number of trainmen in British Columbia had been badly injured or killed by getting their feet caught between the rails and being run down by a train.

On a number of sidings the two sets of parallel rails were too close to provide clearance for trainmen who might be riding on the sides of railcars. These sidings had to be re-laid to provide a minimum of 12 feet between the centerlines of parallel tracks.

The Dolly Varden Mines Railway terminated at the Dolly Varden House, a combined warehouse, office, and residence. On the left, five 20-ton ore cars are spotted at the end of the track. A 19-foot, steel-framed flatcar loaded with supplies for the mine sits in front of the Dolly Varden House. The structure on the right of the Dolly Varden House is the engine house and machine shop. On the dock above and to the right of the barge is the derrick used to lift freight from barges or ships onto the dock. Further to the right, a Porter saddle-tank engine waits to help raise one of the bents into place on the end of the 3000-ton ore bunker. The mainline runs along the hillside to the right and up past the town of Alice Arm seen in the distance.

It was Rae's opinion that many of the curves on the line were too sharp and would require straightening. There was insufficient clearance between the cabs of locomotives and the walls on either side of the rock-cuts. He recommended the construction of a safety switch near the midpoint of the 5½ percent grade between miles 14 and 16. The safety switch was designed so that a runaway train on the upper half of the grade could be diverted by a switchman into the safety switch that ran back uphill and would bring the train to a halt. This safety switch was constructed and is visible on air photos taken in the 1960s.

At the time of Rae's inspection, the two small 12-ton Porter saddle-tank locomotives and engine #3, a 14-ton Davenport 0-4-0 leased from the Granby Company at Anyox, were in use. Rae had grave reservations about the use of this type of locomotive in hauling heavy ore trains and he told the Taylor Company so in his report: "As these locomotives are of the construction contractor's type, they are entirely unsuited for your service." (9.2)

The cars on the railway did not comply with Railway Department requirements as they were equipped with old link-and pin couplers instead of the automatic type. Air-brake equipment was on hand for 13 flatcars and one boxcar but had only been

A loaded ore train rounds the curve past the town of Alice Arm in the summer of 1919. Ole Evindsen took this photo from the grounds of his Alice Arm Hotel.

partially installed. The line's two 4-wheel cabooses were condemned outright. With the cabooses out of service, the boxcar became a *de facto* caboose. It also served as the only passenger conveyance on the line. It had to be equipped with an air gauge and a conductor's valve so that the conductor or brakeman could apply the brakes from inside the car.

In his report, Rae advised Chief Engineer Proctor that the steel was laid on the grade for about 16 miles. A large force of workmen was clearing up slides and filling in washouts along the line. Several bridges still needed decking and it was expected that the railway would be completed to the mine within a month.

Of all the defects noted, the light, 35-pound rail was Rae's chief concern. There was no question that it would not stand up to heavy ore traffic. He suggested the entire line be re-laid with 56-pound rails.

The existence of so many defects and difficulties on the railway made the issuance of a certificate to operate out of the question. The department would allow the railway to be operated for construction purposes only, but not as a carrier of freight until all of its defects had been remedied. This did not stop the Taylor Company from shipping ore as soon as they were able to do so. Nevertheless, the Taylor Mining Company was impressed with the need to clear up the problems revealed in the inspection and set a large force of men to work on bringing the line into compliance with provincial requirements.

At the time of his July visit to Alice Arm, Rae was informed by the Taylor people that they were negotiating for the purchase of a geared locomotive for use on the 5½ percent grade and on the upper end of the line. In August they took delivery of a new 30-ton Climax locomotive that had only recently been "out-shopped" from the builder's plant at Corry, Pennsylvania. The shiny little machine was a fitting demonstration of the Taylor Mining Company's determination to turn the mine and railway into a first-class operation.

In the summer of 1919, a wave of optimism rippled through Alice Arm. The pessimism of the previous winter gave way to an excitement that could only be appreciated by those who had "boomed" the instant mining towns of the Old West or the more recent Klondike Rush. Alice Arm was no Cripple Creek or Dawson City, but in the absence of such bonanzas, it was a fitting substitute. The Dolly Varden scandal had helped put Alice Arm on the map and if the Taylor Mining Company had any say in the matter, they would keep it there.

The left side of Dolly Varden Mines Railway Locomotive #7. This photo was taken at the Climax plant in Corry, Pennsylvania where it was built.

Ore barges float low alongside the Dolly Varden House. The track of the Dolly Varden Railway is temporarily cantilevered over the water so that ore can be dumped directly onto the barges. When one side of a barge was filled, it was turned so the other side could be filled. The barges took the ore to the Granby Consolidated Mining, Smelting and Power Company Smelter at Anyox.

With every new boat in from Vancouver or Prince Rupert there was another group of prospectors, mining engineers, promoters, and speculators who had come to seek their fortunes in the new camp. With the arrival of summer weather there were more men in the mountains surrounding the Kitsault than at any other time in the history of the area. Many of the mining properties were under active development and claims were bringing high prices. The completion of the Dolly Varden Mines Railway would assure the district that the heavy equipment needed on the numerous mining claims could be transported to the railhead at Camp 8 or to intermediate points along the line. From the end of the railway, the government was working on the completion of pack-trails and tote-roads to the mines in the surrounding hills. Alice Arm's future as a "boomtown" seemed secure.

On July 28, 1919, the first trainload of ore was transported over the railway to the wharf at the Dolly Varden House for shipment by barge to Anyox. The Taylor Company was planning the construction of ore bunkers with a capacity of 3000 tons. In the meantime, a temporary ore transfer was arranged at the Dolly Varden wharf by running a short spur onto a cantilevered platform. The ore was dumped through a hole in the platform and deflected onto the ore barge by an inclined ramp. A switch engine spotted the ore cars over the ramp, one at a time, then placed the empties on the wharf track to await their return to the mine.

3276
DOMINION
PHOTO CO.
VANCOUVER B.C.

A general view of the town of Alice Arm from the grounds of the Alice Arm Hotel. There are Dolly Varden Mines Railway ore cars in the right foreground on the storage spur. Space was at a premium along the waterfront and the only way space for the track could be obtained was to build up trestle work over the mud flat.

Left:
Dolly Varden Mines Railway Locomotive #7, the line's Climax locomotive, assembles her train at the foot of the 5½ percent grade near mile 14 for the run down to Alice Arm in the summer of 1920. Because of her braking capacity she had to "double the hill" four cars at a time to get her train of eight cars down the steep grade to this relatively level spot.

With the arrival of the Climax locomotive during August of 1919, it became possible to operate larger trains over the upper parts of the line. Empties were taken up to a siding at the foot of the 5½ percent grade, near mile 14, by one of the small saddle-tank locomotives. From that point the Climax took the empties up the hill to the mine. The saddle tank then returned to the wharf with loaded cars brought down from the mine by the Climax. Ore trains usually consisted of eight cars. The Climax could only pull four empties up the hill, so the first four cars were left at the siding at the upper end while the locomotive returned to the lower end to pick up the rest of its train. This was known as "doubling the hill." From the top of the 5½ percent grade, the full train of eight cars was taken to the main bunkers at Camp 8 for loading. On the way down with loaded cars, the Climax had to reverse the doubling procedure because it had braking capacity for only four loaded cars.

The locomotives of the Dolly Varden Mines Railway were no "shining steeds of the high-iron" with silvered rods, oil and graphite smokeboxes, and polished brass. They were small, squat, stubby "hogs"; devoid of those "higher" attributes possessed by their mainline sisters. Nevertheless, they were possessed of an utilitarian grandeur all their own; they smelled of hot oil and burning coal, and their pace was but a fraction of that possessed by their mainline sisters. But for all of this, they were indispensable to opening the riches of the Dolly Varden Mine and the numerous other mining properties in the Kitsault Valley.

By September the mine was shipping an average of 150 tons of ore daily. Each ton contained an average of 50 ounces of silver. Some of the high-grade material contained hundreds of ounces per ton. The Dolly Varden Mine was attracting the interest of the mining world. It appeared that nothing could stop the run of success that had been initiated by the takeover of the property by the Taylor Mining Company, but there were problems.

On several occasions, the citizens of Alice Arm and the owners of other mining properties in the area petitioned the government to ensure that they also would have access to the Dolly Varden Mines Railway. Although there was no highway as such, a 66-foot-wide strip of land was dedicated to public use. In places the railway occupied a substantial portion of this right-of-way, making it difficult or impossible for wagons or pack trains to pass through. Furthermore, no road would be built because the railway had been given a charter as a common carrier of freight. It had not, however, been granted a permit to carry freight. Nevertheless, it was carrying freight. This was accomplished by the railway retaining its "under construction" status. The Taylor Mining Company could continue construction work on the railway while shipping

The southern end of the railway at the Dolly Varden House. The crew is sorting lumber shipped from the Granby Mill at Larcom Island. One of the 20-ton ore cars, put into service in April of 1920, sits at the end of the line.

ore to tidewater in the otherwise "empty" cars that were returning from the railhead at Camp 8. Using their "under construction" status they could avoid having to ship ore from other mines in the upper valley but they were able to ship their own ore.

The Taylor people however, were having problems with the wharfage at the Dolly Varden House. The Dolly Varden wharf was the only facility of its kind along the entire Alice Arm waterfront. Any freight too large or too heavy to be transferred to shore by lighterage had to be landed at the Dolly Varden wharf and shipped over the railway to the town about three-quarters of a mile away. The delays involved were becoming intolerable to all concerned. In September of 1919, the newly-formed Alice Arm Commercial Club invited the Honorable Dr. J.H. King, Minister of Public Works, to come to Alice Arm to view the wharfage and transportation firsthand. During his visit a meeting was arranged with Alfred Taylor and officials of the Taylor Company, Dr. King, other mine owners, and members of the Commercial Club and the community-at-large. The major concern of the mine owners was that they needed the railway's facilities for carrying supplies and equipment up the valley to their mines and for shipping ore to the smelters at Anyox or farther down the coast. The lack of transportation facilities for these mine operators was greatly impeding the development of the entire Alice Arm area. On several occasions the government had been urged to take over the railway and to increase its rolling stock and facilities in order to

The south end of the Dolly Varden House with two loaded ore cars spotted on the dump track, ready to dump their ore into the barge just beneath.

In the foreground, a partially loaded ore barge has been turned so the empty side nearest the dock can be filled. Oil drums on the other barge contain bunker fuel for the powerhouse at the Dolly Varden Mine.

3290
DOMINION
PHOTO CO.
VANCOUVER B.C.

The middle tower of the aerial bucket tramway that ran from the Dolly Varden Mine to Camp 8. This tramway was 2000 feet long and could transport 12 tons of ore from the upper terminus at the mine to the ore bunker in an hour.

Left:
The upper terminus of the aerial bucket tramway was at the working level of the Dolly Varden Mine. The structure was built entirely of local timber, some of which was cut in the company's small mill at Camp 8.

provide a public transportation service. That, however, was a possible long-term solution to the problem and the mine owners were more concerned about their immediate need for transport.

With regard to shipping the ore of the other mines, Alfred Taylor stated that the availability of railway services for these mines would depend on how well the line stood up under the shipments of ore that had then been underway for only 6 weeks. He assured the meeting that if he was able to do so, he would help the other mine owners and shippers to the best of his ability. In the meantime, work at the Dolly Varden was continuing at an accelerated rate.

By September 20, 1919, a new steam compressor for powering mine drills and auxiliary equipment had been in place and a 2000-foot aerial tramway was in use between the main working level of the mine and the ore bunker at trackside in Camp 8. The tramway was equipped with 1¼ ton buckets and could deliver 12 tons of ore from the upper to the lower bunker per hour. The upper bunker had a capacity of 100 tons and the lower bunker held 150 tons. The tram was fully utilized in shipping as much ore as possible before the end of November and winter with its heavy snows that closed down the railway until spring.

3288
OMINION
PHOTO CO.
VANCOUVER B.C.

Ole Evindsen took this photo of heavy snow on one of the Dolly Varden Mines Railway bridges during the winter of 1919. The bridges were planked over to accommodate the dog sleds that hauled out ore during that winter.

By October 1, the mine had shipped 1000 tons of ore with an average value of $50.00 per ton. In addition, a shipment of high-grade ore from the North Star, which adjoined the Dolly Varden on the north, had been made. This shipment was valued at over $30,000.00 and was the only ore shipped that season not belonging to the Taylor Company. The Dolly Varden was proving to be even richer than had first been believed. The Taylor Mining Company was greatly encouraged by the results of the first season's work and was making ambitious plans for the future.

Additional rolling stock was on order for the railway and some of the cars were received and put into service during November. The other cars, including 20 hopper-bottom wooden ore cars from the Pacific Car and Foundry Company at Seattle, Washington, were not expected until the spring of 1920.

As the days grew shorter and November passed, heavy shipments of ore continued in a drive to ship as much as possible before the snows closed the railway until spring. The snow finally came on December 16 and the last little train of ore cars rattled down to Alice Arm for winter storage. There would be no more railway

Left:
The ore bunker at the lower terminus of the aerial bucket tramway at Camp 8 near the Dolly Varden Mine. The view is from the top of the middle tower. An empty 1¼-ton bucket is being returned to the upper terminus on the cable on the left.

Bridge 16A on the Dolly Varden Mines Railway as seen from the creek bed. The photographer's supply box is at the feet of the two men on the handcar. This trestle, like all those on the Dolly Varden Mines Railway, was built from native wood cut on the site. In the winter of 1919, the tracks on the trestles were planked over so that dog sleds could carry out ore during the months when the railway could not operate because of the snow.

shipments until spring but the Taylor Company pushed on with development work and breaking down ore in the stopes for future shipment.

On December 20, 1919, the Portland Canal Miner, the newspaper voice of Stewart, B.C., carried a lengthy description of the Dolly Varden Mine and Railway under the headline "Snow Blocks Railway and Dog Teams Haul Dolly Ore." It was true, the ores of the Dolly Varden were rich enough to justify the shipment of sacked high-grade ore by dog sled. It was only necessary to plank in the trestle decks on the railway so the dogs could cross over them without mishap.

A SUMMARY OF 1919 OPERATIONS

Under the direction of Maj. Angus Davis, the mine had made a remarkable showing for its first year of operation. In 4 months a total of 6709 tons of ore had been shipped, yielding a total of 423,952 ounces of silver — an average of nearly 63 ounces per ton. In addition, 42 tons of high-grade ore yielding 1280 ounces of silver per ton had been shipped directly to the Tacoma smelter.

3277
OMINION
PHOTO CO
VANCOUVER B.C

10

High-grading the Dolly Varden Mine, 1920 – 1921

During the bitter winter of 1919-20, work continued at the Dolly Varden Mine with a force of about 100 men. Although the Taylor Mining Company was shipping about 2½ tons of high-grade ore by dog sled each day, by the end of December there were 4000 tons of run-of-the-mine ore broken down in the stopes and ready for shipment when the railway reopened. This large stockpile continued to increase daily throughout the winter.

In January of 1920 the company opened a small hydroelectric plant on Trout Creek near the Wolf Mine to generate power for the Dolly Varden Mine and Camp 8. A head of water for the plant was obtained by diverting the waters of Clear Creek through a flume to Trout Creek. A power line was run 2½ miles from that point to Camp 8. The first unit of the plant generated 500 kilowatts and the company planned to increase the capacity to 8000 kilowatts during 1920. During January the company also erected a compressor house at the Wolf Mine and was preparing to continue its work there in the spring.

Sometime before January of 1920 the Dolly Varden miners discovered a vertically inclined body of ore that exceeded their fondest dreams. This "ore chute" was a re-enriched vein 4 feet wide and 75-to-80 percent pure silver. Initially, the Taylor Mining Company tried to keep this discovery a secret but it just wasn't possible in a place as small as Alice Arm. In January of 1920 the *Portland Canal Miner* announced the discovery of this fabulously rich ore at the Dolly Varden Mine. This paper concluded that the discovery of ore so rich would be enough to "tempt Midas to resign his job as universal gold transmuter." Until that time, the silver ores of the Cobalt Mine in Ontario had been regarded as the richest ever found in Canada. The Cobalt Mine had been one of the wonders of the mining world when it was opened in 1905. It had maintained this reputation for 15 years, but the ores of the Dolly Varden Mine were even richer than those of the Cobalt Mine. This time the local experts were sure that this record would never be broken; they felt the ores of the Dolly Varden would remain the richest silver ores in Canada.

Run-of-the-mine ore from the Dolly Varden, shipped to Anyox by barge, yielded 70-to-80 ounces of silver per ton. Before the discovery of the re-enriched chute, a shipment of 21 sacks of what the company had believed to be high-grade ore had yielded $7000.00 worth of silver. These shipments seemed hardly worth the effort when compared with shipments of the newly-discovered, re-enriched ore. By mid-January of 1920, 400 sacks weighing 100 pounds each were ready to ship. These 20 tons of ore were valued at more than $500,000.00. In preparing this shipment, two men in one day had sacked over $10,000.00 worth of ore. Adding this new production to the ore already shipped by the Taylor Mining Company, the silver production of the

Left:
On the Dolly Varden Mines Railway, the steep curving grade (partly 5½ percent) between miles 14 and 16 could be extremely slippery during the frequent rains in the upper Kitsault River Valley. This stretch of track was the site of the derailment of locomotive #5 during the first week of August of 1920.

The Dolly Varden Mine powerhouse and ore tipple at Camp 8. The ramp down from the tipple leads to the sacking shed where high-grade ore, sorted by hand, was sacked for shipment to the refinery.

Dolly Varden Mine from July of 1919 to January of 1920 was well over one million ounces. Soon news of the success of the Taylor Company's efforts reached the outside world and the fame of the Dolly Varden was established.

Thus it came as no surprise to the Taylor Company management when the mine's former owners made an attempt to re-establish their claim to the properties. Even though the Dolly Varden Mines Company had been managed as an investment game it was still very much alive. The owners saw a way of possibly recouping their losses. In January of 1920 the Dolly Varden Mines Company of Chicago notified the Department of Justice in Ottawa that they planned to challenge the Taylor Mining Company's title to the Dolly Varden properties. The title to the properties was a matter solely within the jurisdiction of the Province of British Columbia. The Provincial Government made it clear that if the federal government in Ottawa were to overrule the decision made by its select committee a year earlier, it would re-enact the legislation of the Dolly Varden Mines Railway Amendment Act, 1919, in a way similar to what it had done with at least one other piece of legislation that had been disallowed by Ottawa. Nevertheless, in February of 1920, the Canadian Federal Department of Justice notified the Taylor Mining Company that it must show cause why their title to the mine and railway should not cease. The Dolly Varden Mines Company was given until April 11, 1920, to appeal to the Canadian federal government in Ottawa to overturn the select committee's decision of the preceding year.

Premier Oliver responded to the Canadian federal challenge by introducing a piece of legislation, An Act Respecting the Taylor Engineering Company, confirming their title to the properties. The Taylor Mining Company, established in May of 1919 by the Taylor Engineering Company and its creditors, was the beneficiary of this act. This act confirmed the Taylor Mining Company's ownership of the properties subject to its continued liability for any debts not discharged and remaining among those obligations it had assumed a year earlier. This act received a second reading on April 17, 1920, and was the center of quite a sensation.

During the debate a member of the House rose to read a telegram from R. T. Elliott, counsel for the Dolly Varden Mines Company. This telegram accused Premier Oliver of having broken faith (because he had promised there would be no more Dolly Varden legislation) and of deliberately withholding information from the legislature on the Dolly Varden problem. Elliott also accused Premier Oliver of having deliberately deceived the legislature by pressure of secret lobbying.

A resolution was placed before the Speaker of the House asking that Elliott be called before the Bar of the House to answer for the accusations he had made in the telegram. Premier Oliver saw that little would be accomplished by such an action, and encouraged defeat of the resolution. It was defeated and the Premier indicated he would deal with Elliott's charges in his own time and in his own way. He also said "this constitutes one of the strongest statements of alleged fraud and the using of money for corrupt purposes ever sent to a premier of any province, and I want to say right here and now that I don't intend to stand for any such thing from any living man." (10.1) In spite of these sensational events, the bill passed on the same day that the Elliott

Miners emerging from Dolly Varden Mine tunnel #4 with a load of high-grade ore. The cars were pushed over to the upper tramway and dumped into the bunkers there. This ore was 75-to-80 percent pure silver and was some of the richest ever found in British Columbia. This rich ore did not need to be refined; it was sacked and shipped directly to the smelter.

These three buildings on the main street of Alice Arm were consumed by fire in 1930. Shown here in 1920 are the Kitsault House, one of the town's three hotels, the Kitsault Cafe, and the F. Bruggy Grocery Store.

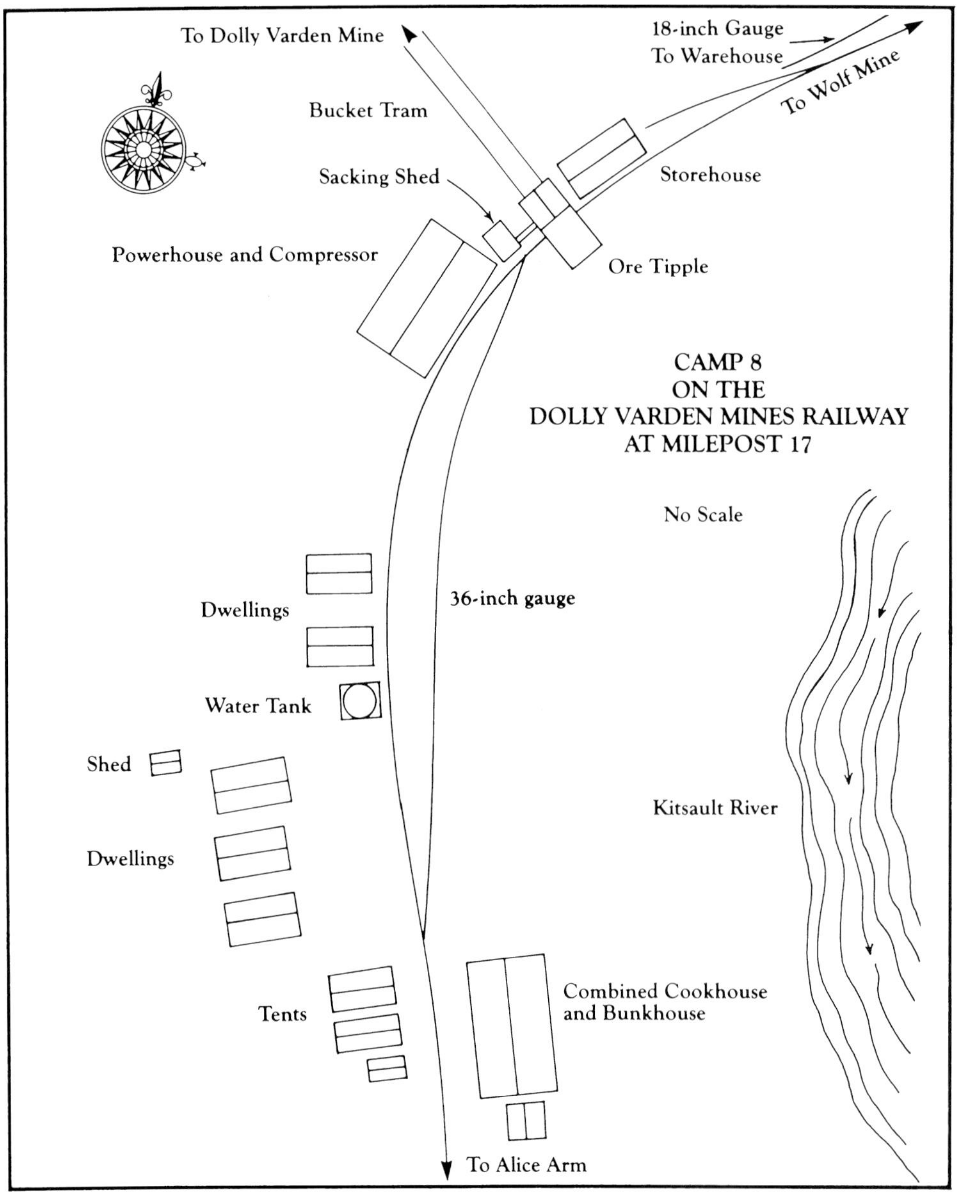

Right:
Dolly Varden Mines Railway #8, a Baldwin 2-4-2T is shown here as operated by her original owner, the Ferries and Cliff House Railway of San Francisco in the 1890s. She was sold to the firm of Armstrong and Morrison of Vancouver, B.C. in 1906 and re-boilered by the Albion Machine Works before being sold to the Clayburn Company of Abbotsford, B.C. The Dolly Varden Mines Railway purchased her from the Clayburn Company in 1920.

telegram was read, and the title of the Taylor Mining Company to the Dolly Varden properties seemed assured. While these matters were being dealt with in Victoria, the work at the mine continued without interruption.

The year of 1920 was the busiest year ever in the little boomtown of Alice Arm. New houses and businesses were springing up and there were plans afoot to open a new townsite on the hill west of Alice Arm. Three new hotels were planned or under construction, but none could compare with Ole Evindsen's Alice Arm Hotel, built back in 1916. Ole's establishment occupied the best site in the area and had a commanding view of the harbor and its approaches. The Alice Arm Hotel had about 20 guest rooms, and a bar and dining room. It was one of the legendary "watering holes" of the northwest coast and no trip north was complete without a night in its comfortable accommodations.

THE RAILWAY – MARCH 1920

William Rae, Chief Inspector of Equipment for the Department of Railways, visited Alice Arm in March of 1920 to examine the railway rolling stock and locomotives being overhauled in preparation for the resumption of service in May. He drafted a set of "Provisional Operating Rules" containing no less than 50 specific prohibitions or instructions to govern the safe operation of the Dolly Varden Mines Railway and to ensure the protection of its crews. Most of the other railways under the jurisdiction of the Department of Railways were standard gauge, 4-foot 8½ inches, so the rules were modified to fit this 36-inch narrow gauge line. For example, the railway's operating rule #50 stated that no trains of more than five loaded cars were to be taken down the 5½ percent grade on slippery rail, and that the conductor was to use his discretion in taking down even that many cars.

The freight station at mile 1 on the Dolly Varden Mines Railway. Number 5 pauses during her chores to pose for the photographer on a misty summer morning in 1920.

In his report to Chief Engineer Proctor, Rae noted that the Taylor Company had purchased two secondhand locomotives. These locomotives came from the Clayburn Company in Abbotsford (located in the Fraser Valley near Vancouver almost on the Canadian-U.S.A. border). The Clayburn Company had recently converted its narrow gauge, clay-mining railway to electric operation and no longer required these two steamers. They were overhauled at the Vancouver Engineering Works before shipment to Alice Arm.

Rae discussed the condition of the Dolly Varden Mines Railway and its equipment with Chief Engineer Sawford of the Taylor Mining Company. In spite of Rae's recommendations of the preceding summer, many of the cars were still equipped with link-and-pin couplers. This type of coupler was considered unsafe by the Department of Railways because trainmen had to step between the railway cars in order to couple and uncouple them. Many had been maimed, losing fingers or hands when the cars crashed together unexpectedly. Some had been killed when they were crushed between cars. Rae and Sawford agreed that the link-and-pin couplers would have be replaced.

The Taylor Company had 20 new ore cars on order from the Pacific Car and Foundry Company of Renton, Washington. They were equipped with automatic couplers and complied in every way with the Department of Railways' requirements. Even the earlier recommendation that the line be re-laid with heavier rail was being complied with, and 56-pound steel was put down in place of the old secondhand, 35-pound rail.

With these improvements the Dolly Varden Mines Railway was ready to operate with increased safety and efficiency. It appeared that all was in readiness for an early resumption of operations, but one more obstacle was put in the way on April 22 when the miners went on strike.

Right:
In this photo taken in the days before stop-action cameras, ore flows from the bunker into one of the Taylor Mining Company's new 20-ton ore cars. These cars purchased in early 1920 from the Pacific Car and Foundry Company had automatic couplers and met other requirements of Inspector Rae of the British Columbia Department of Railways.

3286
DOMINION
PHOTO CO.

3285
PHOTO CO

THE MINERS' STRIKE

Above:
A view down the high line spur toward the town of Alice Arm. Dolly Varden Mines Railway Locomotive #9, a Brooks Mogul, waits for a call to service. A small child sleeps on the porch at the left, unaware of the frantic activity at this busy terminus of the railway.

Left:
A bucket of ore enters the tipple from the Dolly Varden Mine workings. It must be late in the day because the sacking shed is piled high with sacks of the fabulous high-grade Dolly Varden silver ore. One of the railway's condemned 4-wheeled cabooses is just barely visible under the end of the porch beyond the tipple structure.

By April of 1920, a strike was just about the only calamity that had not befallen the Taylor Mining Company since its takeover of the Dolly Varden properties. The preceding year, Rae had refused to issue a certificate to allow the road to operate except for construction purposes because of assorted defects and difficulties. He had condemned the line's only two cabooses. Then, after expensive repairs and new equipment purchases, the line had survived a bitter winter only to have the Taylor Company's right to its title questioned. All of this was followed by the strike at the worst possible time for the owners – the summer operating season. The miners well knew how anxious the company was to ship the large quantity of ore that was broken down and waiting shipment in the mine bunkers.

The miners demanded an increase in wages from a base rate of $5.00 per day to $6.00 per day. They wanted single cot bunks, free blankets and sheets with no increase in the company's rate for room and board of $1.25 per day. Initially, these demands were too much for the company to accept so it decided to break the strike by hiring more miners. Men were hired in Vancouver and Prince Rupert and sent north on the first boat possible. When they arrived at Alice Arm and saw the local mining situation, they unanimously refused to work and joined the strikers instead. The Taylor Company could not afford to have its operation suspended for more than a very short time, and by mid-May it was forced to settle with the workmen.The miners' wages rose to $6.25 per day, muckers' wages rose to $5.75 per day, and trackmen on the railway would now receive $5.25 per day. Union scale was to be paid to cooks and other employees not separately listed. With the strike settled, the 1920 summer operations were finally ready to begin.

By July 1 the railway was in operation and barge shipments of ore were being sent to the Granby Company's smelter at Anyox. Construction was under way on new

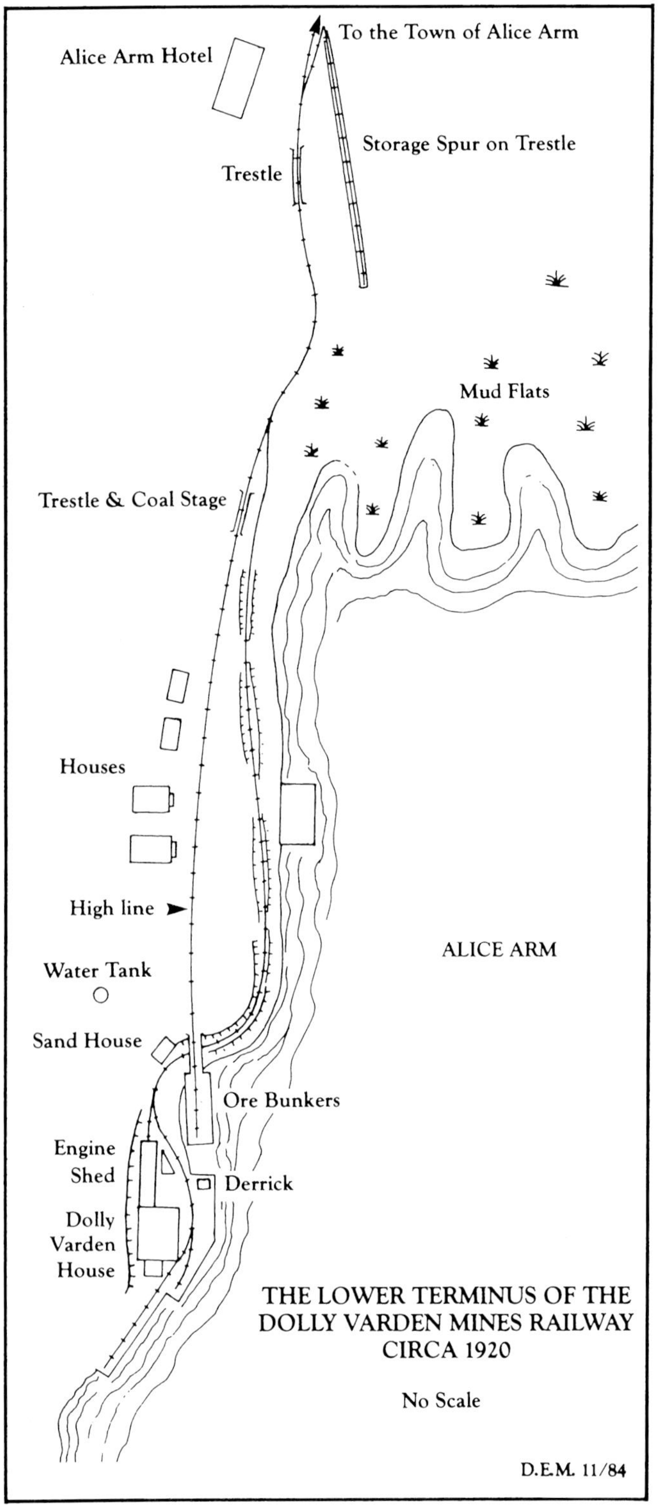

Right:
Ole Evindsen took this photo in Alice Arm on July 1, 1918. A procession of citizens marches down the main street to the sports field. In that year work was proceeding on the Dolly Varden Mines Railway but no ore was shipped until July of 1919.

The Dolly Varden Mines Railway ore bunkers under construction. The view is from the end of the high line railway spur.

3000-ton ore bunkers at the Dolly Varden House. A high line was constructed so that ore cars could be positioned over the bins for dumping. The new automatic coupler rolling stock, ordered a few months earlier, was on hand. The mine and railway were operating with greater shipping capacity than ever before; in the month of July over 5000 tons of ore were shipped. In August, shipments were even greater but were temporarily halted when the first serious accident occurred on the railway.

A RAILWAY ACCIDENT

During the first week of August, 1920, Engineer McCauley, a new man, was given charge of locomotive #5 for a trip from Camp 8 to the bottom of the 5½ percent grade. At that point the train was to be handed over to a locomotive coming up from the terminus at the Dolly Varden House. Conductor Dunn was in charge of McCauley's train. The two decided that #5 could handle two loaded ore cars and a flatcar loaded with passengers and some blasting powder. With crew and passengers on board, McCauley carried out his brake test, cracked the throttle and began his run down the line.

These two ore cars overturned in an accident on the Dolly Varden Mines Railway in August of 1920. Engineer McCauley, new on the line, had decided that the diminutive saddle-tank locomotive #5 could handle two loaded ore cars, and a flatcar loaded with passengers and blasting powder on the 5½ percent grade near mile 15. The train became a runaway when it crested a hill, started to gather speed, and failed to respond to the brakes. In the ensuing investigation, the Inspector found that beyond McCauley's error in judgement, there were serious deficiencies in maintenance and construction that contributed to the accident.

Dolly Varden Mines Railway #5, a Porter saddle-tank engine, stands by to hoist another bent into place on the ore dock. The timber in the water has been brought from the Granby Sawmill at Larcom Island for use in the construction. Number 5 was later involved in an accident on the 5½ percent grade near mile 15 and was declared too small to haul heavy ore trains on steep grades.

The train was under control until it passed a siding at the top of the 5½ percent grade. After cresting the hill it suddenly began to gather speed. Dunn asked McCauley to apply the brakes, McCauley complied, with no result. A second brake application was made with no effect and the speed continued to increase. Dunn ordered the passengers to jump and followed them in company with McCauley.

The un-manned locomotive careened downgrade for another 100 yards then jumped the track and landed on her side with the cars piled around her. Fortunately, the powder on the flatcar did not explode. The matter was reported to a representative of the British Columbia Workmen's Compensation Board who happened to be in Alice Arm. He notified the Department of Railways, and Chief Inspector of Equipment Rae dispatched Inspector A. McPhee to investigate.

McPhee's examination of the wreckage showed that while the brakes had worked, their efficiency was greatly reduced by oil on the brakeshoes and on the wheels of the locomotive and cars. The journal boxes were supposed to hold the ends of the axles in baths of oil and packing material. McPhee found that they had been packed carelessly and there appeared to be more oil on the outside of the journal boxes than within. This was a very dangerous situation, but that wasn't all.

When McPhee investigated the scene of the accident, he found five consecutive rail joints secured with only one bolt in the end of each rail. The other holes in the rail were blind and needed to be drilled so bolts could be threaded through. The rail and the joining plates were secondhand materials from several sources and were not compatible with each other. When the crew was drilling new holes to connect them properly, they had not finished their job and left some holes partially drilled. This problem was well on its way to solution by the time McPhee found it since the Taylor Company had already begun to re-lay the 5½ percent grade with the new, recently arrived, 56-pound rail. There was, however, one other factor that contributed to the accident.

Since Engineer McCauley was new on the line, he had never run a train over it before the day of the accident. According to standard railway practice, he should have had a "pilot," an engineer with experience on the line, to guide him on his first trip in a manner similar to the way an experienced pilot guides the captain of a ship in an unfamiliar harbor.

Inspector McPhee was accustomed to working on large, standard gauge railways and was genuinely shocked at what he saw on the Dolly Varden Mines Railway. He told Chief Inspector Rae that the conditions on the line were deplorable and immediate steps should be taken to keep the line from using the small saddle-tank 0-4-0 locomotives on steep grades with heavy ore loads. McPhee ordered a proper system of routine maintenance and car repair was to be established immediately. In addition, he ordered all trains on the line be under the control of a single dispatcher or company official. Apparently the provisional operating rules drafted by Rae a few months earlier were not being complied with. The Railway Department ordered all deficiencies be rectified immediately. The Minister of Railways had not yet sanctioned the operation of the railway as a common carrier and would not issue an operating certificate until department requirements were complied with.

It was becoming clear to the officials of the Taylor Mining Company that they had to give more attention to the railway, the key to developing the Dolly Varden and Wolf Mines. They began to appreciate the fact that if the railway were given an operating permit as a common carrier, it would generate revenues to help defray the the costs of bringing it up to standard. Since there were eight or nine other mines in the upper Kitsault River Valley that could provide traffic for the line almost immediately, the Taylor Company officials determined to spend additional funds to qualify the line for Railway Department approval. Concurrently, the company management initiated two other expensive projects necessary to the economical operation of the company properties.

A $65,000.00 contract was let for the erection of a new and larger hydroelectric power plant on the Wolf properties to replace the plant that had operated during the previous winter. The new plant would develop enough electricity to power the operation at both mines and have enough surplus capacity to sell power to the town of Alice Arm.

The Taylor Company also let a contract for the first mile of work on the extension of the railway to the Wolf group of claims. The total distance was 2½ miles and the company was also planning to let contracts for the last 1½ miles of work. Under the direction of Maj. C. B. North, a railway engineer, and Maj. Angus Davis, Superintendent of Mines for the Taylor Company, the mine and railway had every expectation for a prosperous future.

The only factor that possibly would affect the construction of the power plant and the improvement and extension of the railway was the price of silver. A fall in silver prices due to a recession in the mining industry would cause a shortage of operating capital. Nevertheless, the resources of the mining properties held by the Taylor Mining Company were large enough to enable it to continue operations long after other companies owning only lower-grade ores would be been forced to quit.

Estimates of the amount of ore at the Dolly Varden and Wolf Mines had been made 18 months earlier by a representative of the Temiskaming Mining Company. The Temiskaming Company was then considering an option on the two properties, and their representative estimated that there were at least 40,000 tons of ore worth 40 dollars per ton at the Dolly Varden. At the Wolf, he estimated that there were 120,000

The ore bunkers near the Dolly Varden House under construction in the summer of 1920. At this time the Dolly Varden wharf in the distance was handling all the freight for the town of Alice Arm, as well as all the supplies for the railway and mine. Because of the congestion the townspeople and mine owners were pressing the Canadian federal government for a public wharf at Alice Arm.

tons at 15 ounces per ton of silver, and 430,000 tons of lower-grade ore suitable for concentration. These low-grade ore bodies were high in silica and could be profitably smelted at Anyox where they could help alleviate the shortage of fluxing material for processing ores from the Granby Company's Hidden Creek Copper Mine. These figures were estimates only, and also did not include the ores proven-up on the other large mining properties in the Alice Arm-Kitsault area. Among these were the Toric, Tiger, Moose, Silver Hoard, Climax, and North Star properties. Without the Dolly Varden Mines Railway, none of these properties could be developed and operated profitably.

The completion of the railway as far as the Dolly Varden Mine and the operation of the mine had generated great excitement among the citizens of Alice Arm and had, so to speak, "put the town on the map." The townspeople and the owners of other mines looked forward to sustaining or increasing the level of prosperity they were then enjoying. They continued to press the Canadian federal government for a public wharf at Alice Arm and the Provincial Government for a road to connect the wharf with the town. The congestion at the Dolly Varden wharf was causing grave difficulty for all concerned, as the wharf was still handling freight for the town. The railway terminus had been constructed on a narrow shelf cut from the sidehill, and the space was so restricted that it was necessary to build a storage spur on trestle work over a

The main street of Alice Arm on a wet morning in 1920.

portion of the tidal flat adjoining the railyard to store the 20 new ore cars that had just arrived from the Pacific Car and Foundry Company.

In the summer and fall of 1920, 150 (or more) tons of ore per day were being moved down from the mines to Alice Arm. These high-volume shipments were interrupted for several days in August when a fierce 4-day rainstorm struck the northwest coast area. It caused widespread damage and the Portland Canal Short Line, 65 miles north of Alice Arm, was left in ruins. The Short Line had only recently been rebuilt and restored to service. Four new bridges on that line were completely destroyed and portions of the grade were washed out. The Portland Canal Short Line did not see operation again until 1929. The Dolly Varden Mines Railway was more fortunate and, while there were slides, washouts, and debris piled on the grade in a number of places, it was returned to service within a few days after the storm was over.

With the railway and mine in operation, there were over 250 men employed by the Taylor Mining Company. Large shipments of run-of-the-mine ore were being made to Anyox daily and smaller shipments of high-grade ore were going directly to the smelter at Tacoma, Washington. The Taylor Mining Company's plan was to reduce the size of the work force to about 65 men after the snows closed the railway for the winter. These men would be employed mining the high-grade ore and on diamond-drilling operations and other work on development of the mine until the railway could be re-opened in the spring of 1921.

The months of October and November passed quickly. In early December the Taylor Company began to lay off some of the workers and to slow down production in preparation for the winter closure of the railway. The Taylor Company planned to continue shipment of high-grade ore during the winter. Al Falconer, a citizen of Alice Arm, was given a contract to ship ore and supplies over the railway grade by horse-drawn sleds.

In late November Falconer began to plank the decks of the bridges along the railway in preparation for the winter shipping season. He started this work at Spring Camp, near mile 9 on the line. The 13,000 board feet of lumber used for this work

The office of the Taylor Mining Company at Spring Camp, near mile 9 on the Dolly Varden Mines Railway.

were cut by Falconer in his own small sawmill. Falconer was one of those astute men who see great mining booms as an opportunity to succeed in pursuits other than mining *per se*. He also operated a general store and later was a contractor engaged in logging operations in the Alice Arm area. (In 1981, Falconer's general store remained as the largest single building in Alice Arm.)

SUMMARY OF 1920 OPERATIONS

During the July-to-December operating months of 1920 the Taylor Mining Company shipped a total of 28,700 tons of silver ore yielding 870,000 ounces of silver. At the end of that time the railway had been improved enough to qualify for an operating certificate from the Department of Railways in Victoria:

CERTIFICATE

In the matter of the Dolly Varden Mines Railway Act; and in the matter of the Taylor Engineering Company: I the undersigned, hereby certify that the railway known as the Dolly Varden Mines Railway, extending from a point being the ore bunkers of the Dolly Varden Mine, 16.62 miles up the Kitsault River from its mouth, thence, following the Kitsault River downstream, to the Public Highway which was described and gazetted on pages 52 and 53 of the "British Columbia Gazette" of the date of the 13th day of January 1916, thence continuing on the said Public Highway down the Kitsault Valley to Alice Arm at or near the surveyed Lot 3635 Cassiar District, is complete and ready for operation.

Dated at Victoria in the
Province of British Columbia

(signed)
A. F. Proctor
Chief Engineer of Railways (10.2)

In order to qualify for this operating certificate, the railroad had undergone major repairs and revisions. The first mile of track from the dock and the 5½ percent grade at mile 14 had been re-laid with 56-pound rail. Over half of the line had been ballasted and nearly a mile of grading was completed on the extension of the railway to the Wolf Mine. Additional rolling stock had been added that complied with the necessary safety requirements.

These repairs and revisions were not the only work on the railway completed in 1920. Large ore bunkers with a capacity for up to 3000 tons of ore were erected at the tidewater terminus and a high line railway track had been constructed so that the ore cars could be run out over the bunkers for dumping. New ore bunkers were also

The completed Dolly Varden Mines Railway ore bunkers sometime in the early 1920s. The Dolly Varden House is on the left with the motor launch, *Awake*, alongside. *Awake* provided passenger and mail service between Alice Arm and Anyox. It's low tide; the high-water line under the ore chutes shows the extreme variation in the height of the tides in Alice Arm.

The switch to the high line. It's high tide in the harbor and in the distance, the launch *Awake* pauses by the boom opposite the Dolly Varden wharf.

constructed at the mine and trackage was laid to them so that equipment could be brought in for the opening of a new tunnel at the 600-foot level. The 600-foot level tunnel was to be driven straight into the mountain to intersect the main ore body. This tunnel would connect with those above by inclined raises so that all ore could be sent directly down to the 600-foot level and dumped in the new bunkers. In addition to all this work, the winter's work of high-grade mining and exploration had been planned.

THE TAYLOR MINING COMPANY'S FINANCIAL CRISIS

Despite the richness of the Dolly Varden ore and the volume and value of its ore shipments, possession of the Dolly Varden properties did not bring instant wealth to the Taylor Mining Company. The financial obligations assumed at the time of the company's takeover of the properties and further developmental costs in putting the mine into production had placed the company in a difficult financial position.

Consequently, while the mine was setting production records and inspiring the awe of people in the mining industry with the richness of its high-grade ores, the directors of the Taylor Mining Company were trying desperately to find additional capital to keep the firm solvent until the assumed debts and costs related to development could be paid off and the mine set up for sustained production of milling-grade ore. The following figure illustrates the company's position well enough:

1. Costs Assumed on Taking Over the Dolly Varden Properties:	
a. Assumption of 1918 costs and profit	$462,638
b. Payment of outstanding wage claims for 1918	134,500
c. Payment of 1919 wages at $5.00 per day 400 men for 100 days (estimate)	200,000
d. Pay off Dolly Varden Mines Company for its investment in the mines and railway	613,000
Sub Total	$1,410,138
2. Additional Costs, Mine and Railway Development, 1920	
a. Contract for construction of hydroelectric plant	$65,000
b. Purchase of 20 ore cars, Pacific Car and Foundry	36,000
c. Payment for 1920 wages at $5.50 per day 250 men for 100 days (estimate)	55,000
d. Erection of ore bunkers at Alice Arm (estimate)	30,000
e. Construction of mine buildings and plant	unknown
f. Development work at Dolly Varden and Wolf Mines	unknown
g. Grading on extension to the Wolf Mine (estimate)	30,000
h. Purchase of 42-foot motor launch, *Gore*	unknown
i. Re-ballasting and refurbishing the railway	unknown
Sub Total excluding unknown costs	$216,000
3. Total Costs per 1 and 2 Above	
a. Costs assumed on taking over the mines and railway	$1,410,138
b. Additional costs 1920	216,000
c. Absolute minimum possible costs to end of 1920	$1,626,138

Silver was valued at more than $1.00 per ounce when the Taylor Mining Company took over the Dolly Varden properties but its value had fallen below that figure by the end of 1920. Using a median value of $1.00 per ounce and the 1,255,590 ounces of silver produced by the mine between July of 1919 and December of 1920, we can estimate mine revenues for this period to be $1,255,590.00 – leaving an operating deficit of $427,048.00. Bearing in mind that this is a minimum possible figure and that there were a number of additional costs, we can see that the financial position of the Taylor Mining Company in December of 1920 was as bad as the position of the Taylor Engineering Company in October of 1918 at the time of its assignment for bankruptcy.

In response to this growing crisis the company decided that Majors North and Davis would plan the 1921 operations while Alfred Taylor and the directors concentrated their efforts on a search for investors with sufficient capital to finance the continuation of work on the mines, railway, and associated facilities. This search ultimately extended to the United States, England, South Africa, and Australia.

Taylor had actually escorted the first of several parties of prospective investors to Alice Arm in May of 1920. They were from the United States, and while they must have been impressed by what they saw, they chose not to invest. Other groups followed during 1920 and 1921, but the result was always the same – silver prices were down because of an oversupply of silver on world markets and there were other investments that promised a better return. Soft silver markets and low prices were threatening to bankrupt the Taylor Company for a second time.

Because the mining firm could not mine more ore or carry out further mine development without more capital it was not going to be able to retire the huge assumed debt in the fixed order of priority directed by the Dolly Varden Mines Railway Amendment Act, 1919. By the end of 1920, George Wingfield of the Goldfield Consolidated Mines Company had received no payment at all against the $150,000.00 (plus accrued interest) the Taylor Engineering Company owed his firm (Provision 3 of the select committee recommendations). This loan originally given to the Dolly Varden Mines Company had been swollen by interest to about $180,000. In spite of the assurance of payment given to him when he withdrew his petition against the Taylor Mining Company in the spring of 1919, Wingfield was still waiting for his money. By April of 1921, his patience was at an end and he moved to institute court proceedings to recover the full amount owed him.

When he was notified of Wingfield's action, Taylor contacted him and negotiations followed. The Taylor Mining Company gave the Goldfield Consolidated Mines Company a mortgage to secure the $182,322.84 owed. Additionally, Wingfield was given a lease that demised the Dolly Varden Mines Railway to him for a period of 999 years. These stringent arrangements secured Wingfield's continuing credit for only another 6 months. This transaction was approved by the Provincial Government under Order-In-Council #347 of March 15, 1921. (10.4) All things considered, it was a brief reprieve, but it gave the Taylor Mining Company 6 months more to raise the capital it so desperately needed.

With these arrangements completed, Taylor left for England where he hoped to interest British capital in the mines and other properties at Alice Arm. Continued control of the Dolly Varden properties by the Taylor Mining Company would depend on the success of his mission. In the meantime, the Wingfield interests were making arrangements with Major North at Alice Arm for the shipment of ore already broken down in the stopes or stored in the Dolly Varden Mine bunkers. The shipment of this ore would permit the Goldfield Consolidated Mines Company to recover at least a portion of the moneys owed it by the Taylor Mining Company.

All of these matters were carried out under tight security and, although there were occasional rumors, the other mine owners and the citizens of Alice Arm were unaware of what was happening or what actions might be taken to operate the Dolly Varden Mines Railway during 1921. Whatever the case might be, they were vitally concerned that the railway should be operated, if not by the Taylor Mining Company, then by the Provincial Government or by one of the other mining companies. The operation or non-operation of the railway would determine the amount of development work carried out at the other mining properties in the Kitsault River Valley.

On May 13, 1921, the *Canadian Mining Journal* described the mining situation at Alice Arm as "not too bright." The silver market was at a low ebb and whether the mines would operate during 1921 was problematical. Nevertheless, the prospectors and mine owners took a long-term view and were certain that the hills along the Kitsault would be the site of feverish prospecting activity just as soon as the melting winter snows would permit. But, further developments on the Dolly Varden properties had to await the return of Alfred Taylor from England.

At the end of May of 1921, Major North returned to Alice Arm from Vancouver to carry out a detailed survey of the Dolly Varden and Wolf properties as a prelude to operations planned for later in the year. The silver market in June of 1921 was anything but encouraging but North pushed ahead with his work. Finally in early July he announced that development work would resume at the Dolly Varden and Wolf Mines. Operations were to be on a reduced scale and the company announced plans to ship only enough ore to cover developmental costs. Rail shipments of ore began

Miners working with a compressed air drill in Dolly Varden Mine Tunnel #4.

and by July 30 the little engines of the Dolly Varden Mines Railway were rolling the ore down to Alice Arm in an almost continuous stream. The first 320 tons of ore were shipped to Anyox by barge at the end of that same week. That first ore shipment was accompanied by 720 cases of blasting powder sold by the Taylor Company to the Granby Company. A stock of 500 cases remained at the Dolly Varden Mine, presumably enough to carry out the year's work there and at the Wolf Mine.

Some stoping work took place at the Dolly Varden Mine but most of the ore shipped consisted of the material left in the mine and on the dumps from the 1920 operations of the Taylor Mining Company. Suddenly on August 13, 1921, North announced that shipments would cease at the end of the following week. He also revealed that the shipments had been made on behalf of the Taylor Company's creditors. This came as an unpleasant surprise to the citizens of Alice Arm; the Taylor Company was in financial trouble once again. It was clear that the railway would not be in operation for the remainder of 1921, but there was consolation in the fact that gas speeders and handcars would be permitted to operate over the line as far as Camp 8. Although the citizens of Alice Arm didn't know it at the time, these speeder operations were a portent of things to come; in future years, speeders were the main mode of transport up the Kitsault River Valley.

Meanwhile, there was growing anxiety over the results of Alfred Taylor's trip to England in search of investment capital. Unfortunately the English visit was a failure. Taylor returned without the money he so badly needed. The continued depression of silver prices and consequent lack of interest in silver mining seemed worldwide – a heavy blow to the Taylor Mining Company and its creditors. After the huge expenditure required to complete the railway and put the mine into production, the prospect of another financial failure was horrible for Alfred Taylor and the company directors to contemplate. For the moment Taylor was the head of the costly empire his faltering company had created around the Alice Arm-Kitsault Valley properties. Sadly, his continued tenure was to be brief. With the expiration of George Wingfield's 6-month reprieve, an action commenced in the Supreme Court of British Columbia to foreclose on the Dolly Varden and Wolf Mines, and on the railway.

To complicate matters, the railway was closed by slides and washouts during a rainstorm in November of 1921. It was impassable even to speeder traffic. In Vancouver, Victoria, and in Alice Arm, the death knell of the Taylor Mining Company was being sounded and many of those directly and indirectly affected were extremely bitter at the unfortunate turn of events.

On December 19, 1921, the *Alice Arm and Anyox Herald* carried a letter to the editor from an anonymous and disgruntled citizen of Alice Arm: "Now, what are the facts? When the Dolly Varden Mining Company applied for a right-of-way from the beach to the Wolf Mine, the citizens of Alice Arm asked the government not to give this company a monopoly of this Railroad. The Dolly Varden Mines Company was made aware of this request and agreed to make of this road a common carrier. They were given a specific time in which to complete the Railroad to the Wolf property and we were sent a wire from our Member, Mr. Mobley, that our rights would be protected. That it was the intention of the old company to build the Railway up to the Wolf is testified by the fact that grading was started and carried on fully one mile beyond the Dolly Varden Mine when the Taylor Engineering Company went broke like they have in nearly all of their undertakings – a favorite of theirs, apparently…" (This writer's attitude appears to have persisted up to the present day with some citizens of the area, but as will be seen, is not an accurate assessment of the facts.)

To complicate matters still further, at the end of 1921 no one seemed to be willing or able to tell the concerned citizens of Alice Arm what was going to happen to either the mines or the railway. The local people were faced with another long North Coast winter and no assurances that "their" railroad would run or "their" mines would operate in the coming spring of 1922.

THE WINGFIELD FORECLOSURE

In May of 1922 with the operating season less than a month away, the *Alice Arm and Anyox Herald* informed its readers that the foreclosure action by George Wingfield to recover his mortgage money and interest was well under way. If the foreclosure was effected immediately, the *Herald* anticipated that it might be possible for the new owners to open the railway and ship ore before the 1922 operating season was over. Besides, there was little to be gained by allowing the properties to lie idle and deteriorate. The railway and mining equipment would soon be rusted, rotten, and useless if they were not properly maintained.

The owners of the other mines in the Kitsault River Valley were, however, more concerned about the railway than they were about the Dolly Varden and Wolf Mines. The Alice Arm chapter of the British Columbia Chamber of Mines, the Alice Arm Commercial Club and other interested parties, requested the assistance of the Provincial Government in having the railway put in condition for the operation of gas speeders and handcars.

In mid-May, A. L. Carruthers, the District Engineer of Public Works for the Southern District of Atlin, arrived at Alice Arm to inspect the railway and to determine the cost to put the line in operating condition. The matter was vital to the citizens of Alice Arm and to the miners as there was no other means of transporting heavy freight and equipment to the upper Kitsault River Valley. Indeed, several mine owners began to put men to work on clearing the lower end of the line of debris. This work ceased by the end of May of 1922 when it was announced that if no mining company cleared the line over the following 2 weeks, the government would do enough work to permit the operation of handcars.

Finally on July 1, the government began the repair work under the direction of J. Anderson. The line was put in shape for light rail traffic and the Alice Arm mining community was able to commence operations for the 1922 season. Meanwhile in Victoria and Vancouver, George Wingfield's foreclosure proceedings were drawing to a close in the Supreme Court of British Columbia.

In addition to Alfred Taylor, the foreclosure action named Hall C. Chiene, trustee for the creditors, and C. M. Ralston, trustee for the bondholders, as defendants. Surprisingly, Taylor, Chiene, and Ralston entered no defense against the action. This led to speculation about a "deal" between the Taylor people and the Wingfield interests. It was rumored that an arrangement had been made whereby the Taylor Mining Company could redeem its interest within a specified time.

On July 7, 1922, the *Canadian Mining Journal* announced that Mr. A. C. DesBrisez, counsel for George Wingfield, had received an *Order Nici* from Justice Morrisson of the Supreme Court of British Columbia for foreclosure against the Dolly Varden and Wolf Mines and the Dolly Varden Mines Railway.

On September 15, the rumors of a deal between the Taylor Mining Company and George Wingfield were confirmed when it was announced that the Taylor Company had been given another 6 months to redeem its obligation to Wingfield. It appeared that although Wingfield had possession of the Dolly Varden properties, he was not anxious to have control of them for very long. The announcement of the additional 6-month extension was greeted with disappointment in Alice Arm. However, the chance of the Taylor firm even trying to obtain the funds necessary to redeem the properties was extremely remote and, after all, the Wingfield interests, in taking over the mines and railway, would have to assume responsibility for any outstanding debts against the properties in just the same way that the Taylor Mining Company had done at the time it took over from the Dolly Varden Mines Company in 1919. With matters as they were in September of 1922, the Taylor Mining Company and its creditors at least stood a chance of breaking even in any resolution of the financial mess surrounding the Dolly Varden properties. Under these circumstances, the Taylor Mining Company and the Taylor Engineering Company were able to bow out of the financial entanglements at Alice Arm and leave the field clear for the Wingfield interests which would hopefully have more financial staying power than the Taylor companies had been able to muster.

3275

11

Dream Mines Die Hard

The George Wingfield interests were infinitely better funded than either of the two preceding holders of the Dolly Varden Mine and Railway. Goldfield Consolidated Mines Company was financially solid compared to either the Dolly Varden Mines Company or the Taylor Engineering Company and its offshoot the Taylor Mines Company. Nevertheless, on Wingfield's takeover of the properties, the debt he assumed was large enough to require him to institute an immediate financial reorganization of the assets he had acquired.

The reorganized company was known as Northern Mining Properties, Limited. It was incorporated in the fall of 1922 and immediately issued shares to the stockholders and creditors of the defunct Taylor Mining Company. The stock certificates were signed by A. C. DesBrisez, attorney for George Wingfield, and by John Anderson, former secretary of the Taylor Mining Company. In fact the new firm was jointly controlled by George Wingfield and the principal stockholders of the Taylor Mining Company.

While Alfred Taylor and his companies were becoming more remotely connected with the Dolly Varden properties, the citizens of Alice Arm still considered Taylor responsible for the financial debacle. (Some of the citizens of Alice Arm interviewed for this book still maintain that Taylor, the Taylor Engineering Company, and the Taylor Mining Company improperly managed the mines and railway.)

On January 27, 1923, the *Alice Arm and Anyox Herald* stated that:

> ... during the two years mentioned (1919 and 1920), when the highgrade was pouring out in a steady stream, the Taylor Mining Company was spending money recklessly, buying equipment for the mines and railway and undertaking costly construction work, but the day of reckoning came, and in December 1920, the mine closed down and the bright light of the Taylor Mining Company which flared in June 1919 was extinguished and the property which had been a steady shipper was closed.

Although some of the equipment purchased by the Taylor Mining Company turned out to be unsuitable, the company's expenditures were for the most part necessary and were part of the drive to put the Dolly Varden and Wolf mines on a long term paying basis. The purchase of the locomotives numbered 8 and 9 was not a wise investment because #8, like numbers 1 to 6, was a small saddle-tank engine of the type the B.C. Railway Department had condemned as unsuited to heavy mining service. Number 9 was a Brooks 2-6-0 (a Mogul-type locomotive) with a wheel base far too long for the tight curves on the Dolly Varden Mines Railway. Number 9

Left:
On a rock shelf carved from the bank of the Kitsault River, Dolly Varden Mines Railway Locomotive #7 backs past a string of cars loaded with the fabulous ores of the Dolly Varden Mine. This Climax was built as an oil burner but had been converted to burn coal when this photo was taken. Number 7 was the only locomotive purchased new by the Dolly Varden Mines Railway.

3261
DOMINION
PHOTO CO.

All except one of the Dolly Varden Mines Railway locomotives were second-hand. Here is Ferries and Cliff House #8 on a passenger run in San Francisco, California, many years before her arrival at Alice Arm in March of 1920. She was built in 1888 and before becoming DVMR #8, she had worked on the United Railways of San Francisco and the Clayburn Mining Railway near Vancouver.

Left:
The Dolly Varden Mines Railway Engine Shed and the Dolly Varden House as seen from the south end of the ore dock. DVMR locomotive #6 is spotting a pair of ore cars on the dump track. The "lazy-leg" derrick on the left was vital to the transfer of supplies from ships or barges to the wharf.

derailed on one of the curves on its first trip up the line and rolled over. It was stored on the siding at mile 2 for many years before finally being cut up by a scrap dealer in 1937.

The only locomotive on the engine roster adequate to its assigned task was the Climax locomotive #7 that handled trains on the 5½ percent grade and on the upper end of the road. It's a shame that #7 should have been condemned to such a short operating career for, after closure of the railway by the Taylor Mining Company in 1921, the only further service she saw was in the Granby Company railway logging operations between 1924 and 1926 (see Chapter 12).

Most of the other equipment purchased by the Taylor Company in 1920 was necessary for the large-scale mining and milling operations that would have made the mines profitable as high-volume silver producers. Since no work was done on the properties by the Wingfield interests during 1922, it's surprising that the the *Alice Arm and Anyox Herald* did not condemn the Northern Mining Properties as well as the Taylor Mining Company. The newspaper couldn't accuse Northern Mining of spending money recklessly but it could have accused the company of not spending any money at all.

The Northern Mining Properties did little or nothing but issue stock and in January of 1924 the company was reorganized again and emerged as Dolly Varden Properties, Limited. This firm was established by Victor Spencer and Associates of Vancouver who initially leased and later purchased the holdings. This new company returned the original name to the "Dream Mine" but unfortunately did little else. There were rumors of the properties being leased to E. G. Taylor, Alfred Taylor's brother, and to some "New York capitalists," but they were never substantiated. Indeed, the only event of significance during 1924 was the lease of the lower 5 miles of the railway (along with two locomotives and other equipment) to the Granby Company. The Anyox smelter complex had a sawmill at Larcom Island, in Granby Bay, and held large timber tracts at Alice Arm and elsewhere throughout the Observatory Inlet area. In addition to its use of the Dolly Varden Mines Railway for logging purposes, the company established another railway on the Kitsault Flats east of the river mouth. The leased railway equipment was used there also.

Dolly Varden Mines Railway Locomotive #9 on the high line at Alice Arm. This 2-6-0, a Brooks Mogul, had a wheelbase too long for the tight curves on the line. She derailed on her first trip up the line and was stored on the siding at mile 2 until the scrappers arrived in 1937.

While mining development in the Alice Arm area continued at a slow pace, the people of Alice Arm faced 1924 as they had the two previous years. There was no assurance as to whether or not the railway would be usable as far as Camp 8 for speeder and handcar traffic or small gas locomotives. Would the Department of Public Works of British Columbia carry out sufficient work to permit operation of these light-rail vehicles on the line as it had during the two preceding years? No one seemed to know, and this pattern of uncertainty and minimal maintenance continued during most years from the 1920s to the railway's scrapping in the mid-1940s.

With each passing year the Dolly Varden Mines Railway became a more quarrelsome political issue. Rumors in March of 1925 had the mine and railway being taken over by the Belmont-Surf Inlet Mines, Limited and speculation arose as to whether or not this company might operate the railway. Unfortunately for the owners of other mining properties in the area and for the citizens of Alice Arm, the Belmont-Surf Inlet deal was only another rumor; Victor Spencer and his partners remained in control of the properties.

In April of 1925 Premier Oliver made another in a long list of pronouncements concerning the railway and Alice Arm. He reiterated his government's position that it was ready at any time to do everything possible for the development of the Kitsault District. While the railway was at least nominally the property of Victor Spencer and the Dolly Varden Properties, Premier Oliver stated that if any mining company should desire to use the line, ways and means would be found to allow them to do so. If any company wished to buy the line and a dispute arose with the nominal owners, a board of arbitration would be appointed to appraise the line's value and award an equitable settlement.

The government's interest in Alice Arm was demonstrated in July when a group of cabinet ministers arrived at the little community for a tour of inspection. They rode over the railway by speeder as far as Camp 8 and visited the Toric Mine adjoining the Dolly Varden claims. The Toric was then under active development and later became one of the major producers in the Kitsault River Valley. The cabinet ministers appeared to have been very favorably impressed but it was some time before any constructive action was taken by the government to resolve the problems of the Dolly Varden Mines Railway. While the cabinet group's trip over the line was not recorded in great detail, the *Herald* described a typical ride over the road only a few months later:

> The Company speeder which makes a daily trip to Camp Eight left Alice Arm shortly before 8:00 a.m. carrying six passengers and 1,000 pounds of freight. The ride up from Alice Arm over the Dolly Varden Railway on a summer morning such as was the case last Tuesday was most enjoyable. The scenery along the banks of the Kitsault is rugged and magnificent and, for those who desire thrills, their wish can be gratified by gazing to the bottom of a few of the canyons en route . . .

Camp 8 on the Dolly Varden Mines Railway. The photographer was on the top of the ore tipple facing south when he took this obviously contrived picture. Two men gaze wistfully at the pile of high-grade ore on the sacking platform and other members of the crew "freeze" for the camera while trying to appear as busy as possible.

The Toric Mine in the late 1920s or early 1930s. This property was located across the Kitsault River to the east of the Dolly Varden Mine. It became a steady producer and was operated almost continuously from 1947 to 1958.

The railway could be as dangerous as it was spectacular. In September of 1925, a speeder operated by the Consolidated Homestake Mining and Development Company was derailed 2 miles above Alice Arm and threw its crew and passengers down the railway embankment. It rolled over J. Strombeck and T. Tavener on the way down, but both men were lying partly in a hollow and were not seriously injured. E. Ness and J. Trinder were thrown onto the track. The most serious injury was received by Strombeck who was evacuated to the Anyox Hospital with a dislocated wrist and broken arm.

In 1926 the Toric Mine was under lease to the Consolidated Homestake Mining Company; this company operated it as Toric Mines, Limited. Toric Mines carried out extensive development work on the properties and agreed to place the railway in operating condition if the government would pay the costs. The government agreed and the line opened for traffic about May 1, 1926.

Toric Mines began work on a concentrating mill in 1926. During the winter of 1926-1927, one of the company's speeders was rebuilt in Vancouver. With this improvement they were able to begin the 1927 operating season with two speeders, a

4-ton gasoline locomotive, and four new 4-wheeled flat cars. The gasoline locomotive was a Plymouth Model TL-2, builder's number 2320, and was built in May of 1926.

In spite of the demise of heavy operations using steam locomotives and rolling stock, speeder traffic over the line was becoming heavier. Thus there was a need for someone to assume responsibility for traffic if accidents were to be avoided, as evidenced by the following poster:

NOTICE! OPERATION OF SPEEDERS OF TORIC MINES COMPANY LIMITED OVER DOLLY VARDEN RAILROAD

It is to be thoroughly understood that, in future, persons wishing to travel on the speeders or locomotive belonging to the above company must enquire of the Superintendent or speeder driver upon the evening previous to their wishing to travel. The above rule will be strictly enforced and persons refusing to adhere thereto will be refused permission to ride. The company reserves the right to refuse permission at any time to any person to travel on its speeders.

(signed)
A. C. H. Gerhardi
Manager

3294

12

The Railway Logging Years

The final operation of the Dolly Varden Mines Railway by the Taylor Mining Company in May of 1921 marked the last use of steam locomotives and heavy rolling stock by the mining industry at Alice Arm. Most of the equipment for the railway was doomed to rust and rot on the sidetracks at Alice Arm and at several other points along the line. From 1923 until the removal of the railway in 1946-47, operations were carried out with gas speeders, small gas locomotives and 4-wheeled trailers. The only exception to this was the lease and operation of the lower portion of the railway and some of its equipment by the Granby Company to assist in logging operations during the 1924-1926 period. Two of the Dolly Varden locomotives (numbers 6 and 7) and some of the rolling stock were used on a separate logging railway constructed on the Kitsault Flats, on the east side of the Kitsault river mouth.

The Abbotsford Lumber, Mining and Development Company had already established a narrow gauge logging line on the flats between the mouths of the Kitsault and Illiance rivers during the spring of 1923. This company logged a large area on the east end of the flats and down the inlet in the area of Silver City (now Kitsault). Prior to these logging operations the economy of Alice Arm had been almost entirely dependent on mining and its fortunes had risen and fallen with the level of mining and prospecting during the annual snow-free season. The development of logging operations was a welcome addition to the economy of the area and, although the rough-and-tumble loggers didn't always see eye-to-eye with the miners, they did tolerate each other.

Left:
The entrance to Dolly Varden Mine Tunnel #3 in the summer of 1920.

THE GRANBY OPENS LOGGING OPERATIONS

In December of 1923 the Granby Company announced it would open logging operations along the Dolly Varden Mines Railway and on the Kitsault Flats during 1924. In the winter of 1923-24, construction began on an isolated section of narrow gauge track east of the mouth of the river and teams were at work getting out piling to be used for log dumping and booming. Simultaneously, a lease was arranged whereby the Granby Company would rehabilitate and operate two of the Dolly Varden locomotives and a number of cars. The almost new Climax #7 was ideally suited to logging work as was #6, one of the 21-ton Porter tank engines. Some of the flatcars were pressed into service as log cars and the Granby Company purchased parts from the Pacific Car and Foundry Company of Seattle, Washington to construct new logging cars. (See Appendix B)

J. A. Dewar and Company #2, shown here, was the sister to J. A. Dewar and Company #3 that was sold to the Abbotsford Lumber, Mining and Development Company and used on the narrow gauge logging railway at Alice Arm. This company logged a large area on the east end of the Kitsault Flats and in the area of Silver City (now Kitsault) in 1923.

The initial logging operations were under the direction of Don Klanahan. Fellers began their work in the woods on the west side of the Dolly Varden Railway during the spring of 1924. Concurrently, the company overhauled its Larcom Island Sawmill in preparation for the season's cut. Meanwhile the railway was still snowed in and ultimately required the attention of many snow shovelers before logs could be shipped by rail.

Two main logging camps were established, employing up to 120 men. Camp 1 was located on the Kitsault Flats and Camp 2 on the Dolly Varden Mines Railway about 2¾ miles up from the Dolly Varden House. Both camps included a full complement of buildings and equipment. Each camp operated three donkey engines consisting of a two-speed Washington Iron Works yarder and separate loading and rigging engines.

Dolly Varden Mines Railway #6, a Porter 0-4-0T, unloads logs at the dump trestle on the Kitsault Flats, east of the mouth of the Kitsault River. In the mid-1920s, logging added a new dimension to the economy of the Alice Arm area that heretofore had been dependent only on the fortunes of the miners and prospectors.

Logs are loaded with a McLean swing boom at the Granby Consolidated Mining, Smelting and Power Company logging operation on the Kitsault Flats near Alice Arm in 1927. These booms were rigged to spar trees and were powered with donkey engines such as the one in the middle of the photo. Three men are barely visible in the left foreground.

The system employed was called high-lead logging and involved the use of spar trees and lines that lifted the leading end of logs clear of the ground before yarding them to trackside for loading onto rail cars.

On March 22, 1924, the company put men to work shoveling snow off the lower end of the railway so lumber could be taken up to Camp 2. The work proceeded quickly and by March 28 lumber was moving up the line by rail for the erection of camp buildings. Felling and bucking continued and by the end of the month there were 2,500,000 board feet of lumber on the ground. A large trestle also had to be constructed at Camp 2 and would not be finished for some weeks. In the meantime, work on the flats was continuing apace.

The car parts from the Pacific Car and Foundry Company were sent to the Granby shops at Anyox and on March 27 four newly constructed skeleton cars arrived at Alice Arm from the Granby shops at Anyox. Four more followed a month later followed by the same number a few weeks later. Two log cars were constructed in the Dolly Varden shops at Alice Arm but the ten cars built from the PC&F parts were the main elements of the Granby rolling stock. With other leased equipment from the Dolly Varden Railway, the logging operations had plenty of rolling stock.

By May the log-dump trestle on the Kitsault Flats was nearly ready and a second one had been set up at the Dolly Varden Railway terminus. The two camps were employing over 100 men and all the bunkhouses, cookhouses and other facilities were complete.

Finally, on Tuesday, May 6, Climax #7 pulled the first narrow gauge log train down to Alice Arm from Camp 2. Shipments soon averaged around 50,000 board feet per day. Occasionally, logs up to 9 feet in diameter and 32 feet long were sent down on the narrow gauge cars. Pictures of these shipments in the local newspaper showed one of the loggers standing by the car to emphasize the bulk of the huge timber. With such fine timber, mainly spruce and some hemlock, the Granby Sawmill at Larcom Island was in full operation. When Camp 1 came into production, the mill put on another shift and increased the cut even more.

At the end of May, D. McQueen, high-rigger for the Granby Company, put on a show at Camp 1 for visitors from Alice Arm by topping-off a large spruce spar tree. This specimen was 140 feet high and 20 inches in diameter at the cutoff point. About 75 percent of the cut on the flats was spruce of the same high quality. Much of the lumber cut was excellent timber known as "clears" with no knots.

By July Camp 2 had increased production from 70,000 to 100,000 board feet daily with all of the timber going down to the dump at Alice Arm by railway. With an average carload of 5000 feet, the shipments averaged 20 carloads of logs per day. While operations elsewhere on the Dolly Varden Mines Railway were confined to gas speeders, the lower end of the line saw a renewal of heavy steam operations

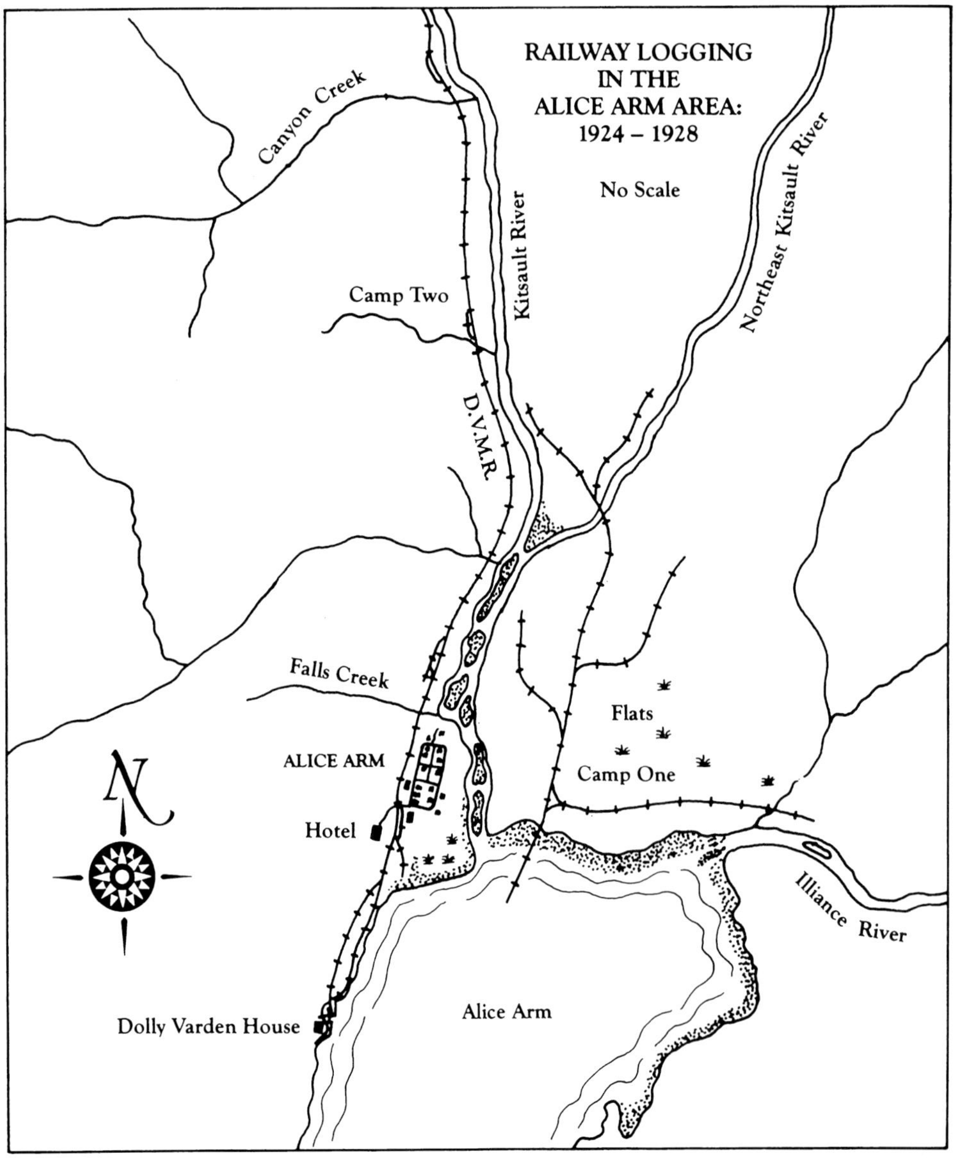

reminiscent of the peak period of operation by the Taylor Mining Company. Soon Camp 1 began rail shipments as well.

By mid-July, tracklaying on the Kitsault Flats was far enough along to permit log shipments and locomotive #7 and four log flats were assigned to this area. Surprisingly, the logging operation on the Dolly Varden Mining Railway closed down for the year at this time.

With the shutdown of Camp 2, the donkeys, locomotive #7 and the logging equipment were transferred to Camp 1. The railway on the Kitsault Flats was extended so it was 4 miles long. With this increase in operations the people of Alice Arm were surprised when cutting was suspended on the flats on September 15, about 2 months earlier than expected. This closure was caused by a sharp downward turn in the lumber market. The citizens learned that resumption of logging in the spring of 1925 would depend on improved market conditions.

The Larcom Island Sawmill had a 2,000,000 board foot surplus of cut lumber in its yards. This inventory would have to be used up in the Granby's Anyox operations or sold before logging in the Alice Arm area would be resumed. In September some railway work was started at Camp 1 but it was soon discontinued. The locomotives and equipment were again laid up to await better times. Unfortunately, 1925 saw a continuation of poor market conditions so no logging operations were attempted for the year. It almost looked as if narrow gauge logging at Alice Arm was over.

In May of 1926 the Climax locomotive and 14 cars were barged back across the harbor to the Dolly Varden House and the Granby skeleton cars and several other pieces of rolling stock were slated to return to Anyox. The local people were surprised when shortly afterward, F. L. Housely of Alice Arm took a contract to reopen logging on the flats for the Granby Company. Housely had apparently convinced the huge Granby Mining Company that he could operate the logging operation more successfully than they could.

A bridge was constructed across the northeast fork of the Kitsault and the railway from Camp 1 was extended into 8,000,000 board feet of fine virgin timber. Housely had an arrangement to sell the timber to the Powell River Company that operated a huge pulp mill at Powell River, B.C., about 90 miles north of Vancouver. If the arrangement proved successful it was even possible that logging could resume at Camp 2.

In October of 1926 the log hulks *Biscayne* and *Bingamon* arrived at Alice Arm under tow by the tug *Lorne* to pick up the first shipment of 3,000,000 feet of timber. Three thousand feet of new railway was completed and 2000 feet of steel had already been laid. The 100-foot-long bridge over the Northeast Kitsault was completed and was the largest structure of its type in the Kitsault River Valley.

In early 1927, Housely shipped another 1,500,000 feet on the *Bigamon* and on the *Basuto*, another hulk. Although these shipments were successful, they were costly. Housely decided to try a more efficient method of shipment: the Davis Raft, a large cigar-shaped bundle of logs held together by cables or massive chains. In March a crew arrived in Alice Arm to construct the first raft and by May two shipments had been made.

HOUSELY SUCCEEDS WHERE GRANBY FAILED

F. L. Housely made a success of the logging on the Kitsault Flats after the Granby Consolidated Mining, Smelting and Power Company had failed to do so. During his year in charge he shipped 8,000,000 feet of spruce and hemlock from the flats and his payrolls and purchases in Alice Arm contributed greatly to the local economy.

Housley kept the camp running through the severe winter period when most other operations were closed. His track gangs built nearly 2 miles of additional railway after he took over, and when he was through the logging had moved from the shoreline to the extreme north end of the flats. When the trees there were all cut, Housely's crews returned the railway equipment to the Dolly Varden House and shifted logging operations down the inlet.

Ole Evindsen photographed the trestle over the mud flat at Alice Arm when it was beginning to collapse in the 1930s. This trestle was west of the mouth of the Kitsault River. Logs were loaded onto hulks from another trestle on these mud flats but east of the river mouth. Evindsen's Alice Arm Hotel is in the background.

The last spar tree was rigged during the final week of August, 1927, and when the last setting was completed, only small, scattered stands of timber remained to be cleaned up. In October the logging railway was pulled up. In December, a hulk carrying a final shipment of logs already brought by rail to the trestles, went to the Powell River Company.

In June of 1928, Al Falconer of Alice Arm secured a contract to lift the last rails and tear out the long dump trestle on the east side of the Kitsault. Shortly afterwards

Ole Evindsen photographed Dolly Varden Mines Railway Locomotive #9. This once proud Brooks Mogul sat forlornly on the siding at mile 2 from the end of shipments from the mine in 1923 until 1937 when it was torn up for scrap.

the camp buildings were removed and shipped out to the Granby Company's mining operations at Quartz Point. The narrow gauge logging era at Alice Arm had come to an end after only 4 years.

With the end of logging the Dolly Varden locomotives and rolling stock rusted away their years until the scrapper arrived at Alice Arm in 1937. Think of the trim little Climax locomotive and her saddle-tank companions, with so much life left in them, condemned to inactivity and a slow lingering death at the hands of the elements. No more would their whistle songs rise among the canyons and wooded hills of the Kitsault River Valley or along the harbor front. With them the romance that had surrounded the glory days of the Dolly Varden Mines Railway was dead.

13

Epilogue: Reflections on the Past

The Dolly Varden Mine story was not finished with the end of the logging years and the scrapping of the locomotives, Victor Spencer and several associates demonstrated their faith in the Dolly Varden mine when they acquired the Dolly Varden claims in 1923 and held them until Spencer's death in the 1950s. After 5 years the Spencer syndicate optioned the properties to the Britannia Mining and Smelting Company, Limited in November of 1929. Under the direction of Britannia some work was done on the main Dolly Varden vein at the west end of the Dolly Varden #2 claim. The Britannia Company also optioned the Toric properties. They formed a subsidiary known as the Torbrit Mines to operate their combined holdings in the Kitsault Valley. However, the option on the Dolly Varden and the other properties was dropped a short time later after another severe decline in silver prices occurred at the beginning of the Great Depression.

There were no further developments at the Dolly Varden Mine until 1935, when T. W. Falconer of Alice Arm secured a lease on the mine. Ore shipments by speeder and 4-wheeled flatcar began in that year and lasted through 1940. Total shipments for this lease period came to only 234 tons that yielded 63,609 ounces of silver, 2047 pounds of lead out of shipments of 222 tons, and 420 pounds of copper. These were the last shipments of ore from the Dolly Varden to date when this book was written in 1983.

TOTAL ORE AND SILVER PRODUCTIONS OF THE DOLLY VARDEN MINE

Period	Ore (tons)	Silver (oz.)
1919	6,709	423,952
1920	28,700	870,000
1921	1,874	45,648
1935-40	234	63,609
Totals	37,517	1,403,209

In 1960, Dolly Varden Mines, Limited was incorporated, and purchased the Dolly Varden Mine from the Spencer estate along with the nearby North Star, Toric, and Wolf properties. The holdings were briefly optioned to the Sunshine Mining Company of Spokane, Washington from January of 1964 to September of 1965, but no work is known to have been carried out.

In February of 1967, Newmont Mining Corporation of Canada optioned the properties and carried out geological mapping, airborne, and ground-magnetic surveys. The option was again dropped and work was resumed by the Dolly Varden

Left:
The Dolly Varden Mine ore bunkers under construction. The planned capacity was 3000 tons of ore.

Mines during the 1967 and 1968 seasons on the Wolf claims. This work was continued in October of 1969 by Bralorne Car-Fer Resources, Limited but the option was dropped on April 30, 1970.

In 1972, the New Jersey Zinc Corporation optioned the mines and carried out surface and underground mapping on the 1638- and 1772-foot levels of the Dolly Varden. Diamond drilling was carried out at five locations on the Dolly Varden #1 and #2 claims and west of the old glory holes. Again, the option was allowed to expire.

Based on diamond drilling reports done in the 1920s, ore reserves today on the Dolly Varden are still 47,443 tons of ore at 16.0 ounces of silver per ton. Combined reserves of the Dolly Varden, North Star, Wolf, and Toric reported as proven, provable, and possible are 1,712,444 tons of ore averaging 9.52 ounces of silver per ton, 0.53 percent lead, and 0.82 percent zinc.

Control of Dolly Varden Mines, Limited was purchased by Royal Agassiz Mines in 1975 and was later passed to the Bulora Corporation, Limited. The properties were reorganized as Silver Dolly Resources in 1975 but failure to meet a payment put them into receivership and the company was put up for sale in 1976. The firm's name was changed to Dolly Resources, Limited in 1976 to Dolly Varden Resources, Limited in 1977, 95-percent-owned by Yorkshire Resources , Limited. Additional development work was carried out on the properties in 1978 and 1980 when they were acquired by Dolly Varden Minerals Inc. In 1985 the owner was still listed by the Department of Mines as Dolly Varden Minerals Inc.

The compelling aspect of the Dolly Varden story is that it's the tale of a mine and railway whose destinies were inseparably linked. *The Dolly Varden Railway existed only because the mine required a cheap means of transportion for its ore. The Dolly Varden Mine came into existence only because there was a railway to serve it.* If the Taylor Mining Company's plans for a concentrating mill to treat the large bodies of low-grade ore at the Dolly Varden Mine had come to fruition, it's likely the railway would have had a longer and more productive life.

Ole Evindsen, one of the discoverers of the Dolly Varden Mine had many occupations in his life. Here he is suited up in preparation for a dive during the construction of the Prince Rupert harbor works in 1906. He owned and managed the Alice Arm Hotel during the years when the railway and the mine brought dreams of riches to the citizens of Alice Arm. In the 1930s Evindsen worked on the Dolly Varden Mines Railway when it was being maintained by the Provincial Department of Public Works. He worked for the U.S. Navy in World War II and died at the age of 91.

Ole Evindsen briefs two young admirers on the operation of a gas speeder prior to a trip over the Dolly Varden Mines Railway in the early 1930s. Gas speeders and 4-wheeled trailers were the only traffic over the upper part of the line after 1923 when the last steam run was made.

There would have been no Dolly Varden story to tell without the efforts of Ole Evindsen and his partners who looked beyond the horizon to find and stake the mine in a remote and little-explored area. After the discovery and early development of the mine, Ole Evindsen's partners vanish from this history; Evindsen does not.

Evindsen was a remarkable man in his own right and a genuine historical figure in northwestern British Columbia. He was born in northern Norway in 1881 and rounded the Horn as cabin boy on a sailing ship at age 14. During the voyage the captain withheld rations from his crew and scurvy broke out. Evindsen avoided contacting the disease because one of the ship's officers smuggled extra food to him.

Upon his arrival in British Columbia, Evindsen joined the crew of a sealing schooner in the Bering Sea and nearly perished when his partner refused to believe the compass and they lost sight of the mother ship. Only a Higher Power intervened and they were rescued. Evindsen refused to ever go to sea with that partner again. Later, Evindsen was a hard-hat diver working on the construction of early harbor works at Prince Rupert, B. C. before joining Pearson, Eik and Carlson in their discovery on the Kitsault River.

The author stood on the old Dolly Varden Mines Railway grade to take this photograph of the town of Alice Arm in August of 1979. The large building is Falconer's store.

The remainder of Evindsen's life was more settled; for some years he operated his Alice Arm Hotel. After selling it he resumed his connection with the Dolly Varden when he became bridge foreman on the railway during the 1930s for the Provincial Department of Public Works. He left Alice Arm for Prince Rupert at the beginning of World War II and after 1942 he worked for the U.S. Navy. Later he settled in Ceepeecee on the west coast of Vancouver Island. He died in Vancouver in 1972 at 91 years of age; his was a useful and memorable life.

The ironies of the Dolly Varden story are twofold. First, as too often happens, the discoverers of the bonanza profited little from it. Second, no one else to date has profited from it either. It's true that the Dolly Varden made some rich shipments but the cost of developing the mine and railway more than consumed the returns on the ore mined.

Had it not been for Hubbard's manipulation of Taylor and the Dolly Varden stockholders this tale might have ended before it had begun. On the surface Hubbard was plainly devious but not truly dishonest. If he had been dishonest he would have denied his commitment to reimburse Taylor before, during and after the Dolly Varden hearing. He made no effort to do so. Hubbard was an enigmatic character, but with many qualities that helped him cultivate the confidence of others. He appears to have been a kindly if somewhat "crooked" man who wanted the Dolly Varden project to succeed for the benefit of all concerned, even if he had to stretch the truth. Those who suffered from his actions included his own backers as well as the workers on the

When the Dolly Varden Mines Railway was scrapped in the late 1930s the scrapper tipped the wooden ore cars off the trestle onto the mudflats and stripped them of their trucks. The remaining parts were left to slowly sink into oblivion. When this photograph was taken by the author in August of 1979, the remains of eight cars were still very much in evidence.

railway and the management of the Taylor Company. Hubbard also apparently deceived himself and this smacks more of one given to wishful thinking than it does of a real scoundrel. He died only a few months after the hearing ended in the spring of 1919.

If the story had been a little different, Alfred Taylor might have gone down as one of the most naive characters in British Columbia mining history. Fortunately, while he was initially taken in by Hubbard he went on to learn from the Dolly Varden affair and continued with a successful career. After Taylor severed his connection with the Dolly Varden properties and Alice Arm he played a role in the development of Vancouver's North Shore. Today, Taylor Heights and Taylor Way commemorate his contribution to the development of North Vancouver.

Taylor's initial lapses in allowing himself to be manipulated by Hubbard were probably due to his belief in the integrity and sincerity of his fellow man rather than any lack of intelligence or ability. He liked Joseph Hubbard and believed in him for a long time. His error in doing so was probably his greatest lesson in his business career. The story of the Dolly Varden Mines Railway and Taylor and Hubbard is sufficient to guarantee them all a place in the history of narrow gauge railroading and mining that would be hard to equal.

After the end of operations the Dolly Varden Mines Railway equipment sat rusting and rotting on the sidings at Alice Arm and along the line through the late 1920s and into the 1930s. The end came in 1937 when a scrap dealer from down the

The preparation of the bridge abutments.

A tripod and standing line with a pulley arrangement are rigged to lift the log into place in the span. These workers had use for strong backs and a lot of ingenuity.

This set of photographs was taken by Ole Evindsen during the re-building of the bridge at Deadman's curve at the entrance to Mud Creek Canyon. This kind of expensive and time consuming maintenance on the railway was one of the reasons that the Dolly Varden Mine never realized a profit.

A pair of trucks from the Dolly Varden Mines Railway's old ore cars was modified with a swiveling saddle arrangement to move heavy logs into place as part of the span of the replacement bridge.

The replacement bridge at Deadman's Curve. The stringers are in place and the ties are spiked down. The crew can move on to clear up the next obstacle further up the line.

In the early 1940s the *Eleanor Mac* and the *Anna H.* stand by at the government wharf at Alice Arm. The *Anna H.* had taken over the mail run from the old *Awake*. In 1940 the Dolly Varden House that once stood on this spot was just a memory as were the silver dreams of the early miners who passed through Alice Arm on their way to their dream mine, the Dolly Varden.

coast came to Alice Arm and cut up the equipment. The wooden cars were turned over and stripped of their trucks and metal parts and the locomotives were reduced by the cutter's torch. It was a sad ending to the dreams of the once proud little railway along the Kitsault. When I visited Alice Arm in August of 1979 the remains of several ore cars were still in the mud flat at the side of the old storage trestle and the old boxcar/caboose still sat on the side of the railway grade several hundred yards away.

The Dolly Varden Mine brought the riches of Northwest Canada to the attention of the outside world. The expense involved in bringing the mine into production and building the railway is a lesson for all of those who look for bonanzas. This story of railway builders completing their difficult work is an inspiration to railway lovers everywhere. We are richer through their efforts and we shall not see their like again.

Appendix A: Equipment of the Dolly Varden Mines Railway

by Darryl Muralt and John Lewis

ORIGINS

The origin of most of the locomotives and equipment of the Dolly Varden Mines Railway is uncertain. The difficulty of obtaining equipment during World War I made it necessary for the Taylor Engineering Company to obtain its engines and rolling stock wherever and whenever it could. The passage of time has further obscured the facts, it is unlikely we will ever know exactly where all of the equipment came from.

In 1914 the Grand Trunk Pacific Railway was completed into Prince Rupert, only 90 miles from Alice Arm, and much of the equipment used on the Dolly Varden Mines Railway during the 1917 construction season apparently came from narrow gauge equipment left over by the builders of the Grand Trunk Pacific. Foley, Welch, and Stewart, the company that built the Grand Trunk Pacific, had purchased a large number of 0-4-0 saddle-tank construction engines including at least 20 from Porter and a smaller number from Davenport and other builders. It is impossible to say which builders built the engines obtained by the Taylor Engineering Company. However, an early Dolly Varden Mines Railway construction contract dated February 23, 1917, describes an engine with 10 x 16 inch cylinders. Cylinders of this size were common on Porter 0-4-0 saddle-tank engines of that day. A Vancouver newspaper article of 1917 also describes the first two engines of the Dolly Varden Mines Railway as Porters.

Left:
Four of the Rogers dump cars used in the construction of the Dolly Varden Mines Railway sit on the lower end of the highline spur at Alice Arm.

INSPECTION REPORTS

The most detailed information available on Dolly Varden Mines Railway equipment is in the files and annual reports of the British Columbia Railway Department. Unfortunately, their first inspection did not take place until October 12, 1918, after the first two locomotives had been disposed of and replaced by a Davenport 0-4-0 saddle-tank engine leased from the Granby Company at Anyox and two secondhand Porters purchased from U.S. mining companies. These "new" locomotives were numbered 3, 5, and 6. No record has been found of a Dolly Varden Mines Railway Locomotive #4. There may not have been one.

This little Davenport was leased from the Granby Company in 1918 and 1919 to run on the Dolly Varden Mines Railway as DVMR #3. She was also Granby #3 and is shown here on the Granby operation at Anyox.

In the report of his inspection, R. W. Macintyre listed the following Dolly Varden Mines Railway equipment:

One 14-ton 0-4-0T (the leased Davenport saddletank, DVMR #3)
Two 21-ton tank 0-4-0T's (the two secondhand Porters both saddletanks, DVMR numbers 5 and 6)
Thirteen 30,000-pound capacity, steel-framed flatcars
Two 4-wheeled, single-truck cabooses
Two 10-foot long by 6-foot-wide flatcars
Twenty-four dump cars (Rogers)
One 47-ton ¾ cubic yard steam shovel
One 20-foot long by 6½-foot wide boxcar

There was a second inspection of the railway in early January of 1919 but no changes were made in the equipment list from the previous October. The locomotives were stored in the engine house on the wharf and the steam shovel was sitting unused on the siding a half mile up the line with its rental fees costing $40.00 per day.

The railway was again inspected in July of 1919 when the Taylor Mining Company took over the Dolly Varden properties.

The Inspector condemned the 4-wheeled cabooses outright because he considered them too dangerous to be used in hauling people. The cabooses were taken off the tracks and at least one of them sat by the ore tipple at Camp 8 where it was probably used as a storage shed.

With the removal of the cabooses from service the boxcar became the *de facto* caboose. It was fitted with side doors and a small door in each end and was put on an all-steel frame. This car sat for many years on the side of the grade at the old yard site in Alice Arm and was still there in the summer of 1979.

The Taylor Mining Company hired Frank Sawford, previously in charge of the Canadian Collieries Railway at Cumberland, on Vancouver Island, to be Chief

Mechanical Engineer on the Dolly Varden line. On July 14, 1919, Sawford wrote to Chief Engineer Proctor of the Department of Railways and informed him that the company was negotiating for the purchase of a geared locomotive for use on the upper end of the line. He also asked the department what weight locomotive would be

Dolly Varden Mines Railway
Porter 0-4-0T

(This plan is for both DVMR #5 and #6)

Traced by Rich Dunn from the original erecting drawing

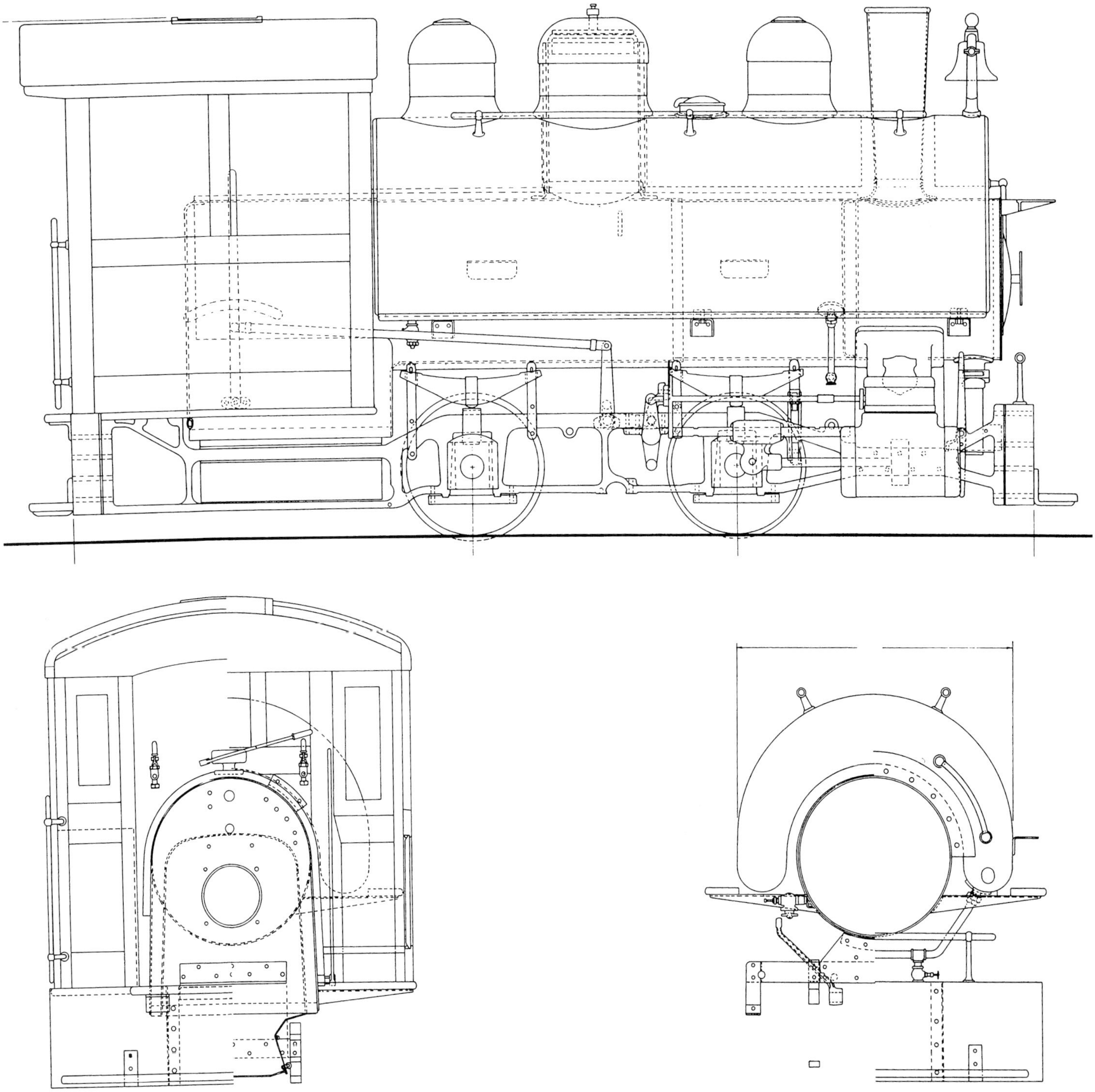

Scale: ⅜ inch = 1 foot

Ole Evindsen photographed this leased steam shovel working in the gravel pit at mile 4 on the Dolly Varden Mines Railway in the spring of 1918.

adequate for the heavy ore trains the company intended to operate. The line had been laid with a mixture of 30- and 40-pound rail, considered by the department to be too light. The department recommended that this rail be replaced with 56-pound rail.

UPGRADING THE RAILWAY

In August of 1919 a new, 30-ton, Climax-geared locomotive was delivered. She became DVMR #7, the only engine purchased new for the operations at Alice Arm. She began service on the upper end of the line at mile 14 and beyond on the 5½ percent grade. The saddle-tank 0-4-0's were then operated mainly from mile 14 to Alice Arm on the less steep part of the railway. However they did go up and down the 5½ percent grade on occasion – but not always safely. DVMR #5, a saddle-tank 0-4-0, was wrecked in the accident described in Chapter 10 when she ran away descending this grade with an ore train and an inexperienced engineer at her controls.

Chief Engineer Sawford and his railway staff were committed to resolving the inadequacies of the railway and its equipment and began a program of improvement. Many of the cars were equipped with air brakes and new 56-pound rail began replacing the lighter rail at the lower end of the line and on the 5½ percent grade. However, Sawford's most memorable contribution to the railway was in his design for the 20 wooden ore cars ordered in the spring of 1920.

The upgrading was carried on through the winter of 1919-1920. In March of 1920 two more secondhand locomotives, Dolly Varden Mines Railway numbers 8 and 9, arrived. A Baldwin 2-4-2T and a Brooks Mogul, they had been purchased in February from the Clayburn Company of Abbotsford, about 60 miles east of Vancouver, B. C. in the Fraser Valley. The Clayburn Company had electrified their line and no longer needed steam engines. These engines had to be rebuilt and spent a month at the Vancouver Engineering works being readied for service on the Dolly Varden Mines Railway.

Dolly Varden Mines Railway #7 at the Climax plant in Corry, Pennsylvania, before delivery to the Dolly Varden Mines Railway at Alice Arm. She was the only locomotive purchased new for the Dolly Varden line.

The left side of Dolly Varden Mines Railway #7 taken at the Climax plant in Corry, Pennsylvania, in 1919.

Dolly Varden Mines Railway #8 as she appeared working for her original owner, the Ferries and Cliff House Railway of San Francisco, California, around 1900.

Dolly Varden Mines Railway #8 during her last year of service for the Clayburn Company, at Abbotsford, B.C. in 1919. She was rebuilt by the Vancouver Engineering Works before being shipped to Alice Arm for service on the Dolly Varden Mines Railway.

Dolly Varden Mines Railway Locomotive #9 after she was rebuilt at the Vancouver Engineering Works. She was completely overhauled before shipment to Alice Arm but was little used because of her long wheelbase and the tight curves on the Dolly Varden line.

Dolly Varden Mines Railway
Climax (DVMR #7)

Drawn by Rich Dunn

Scale: ¼ inch = 1 foot

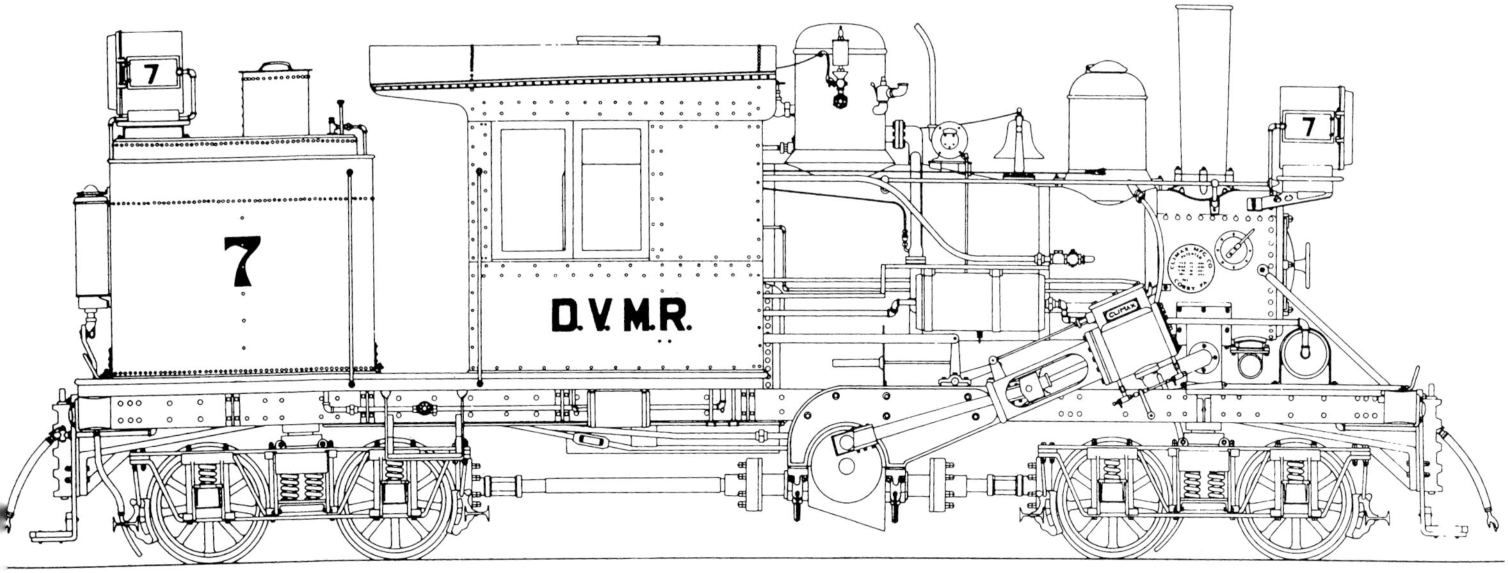

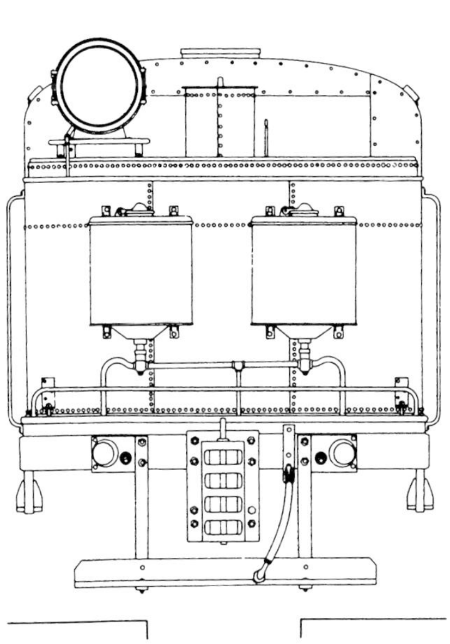

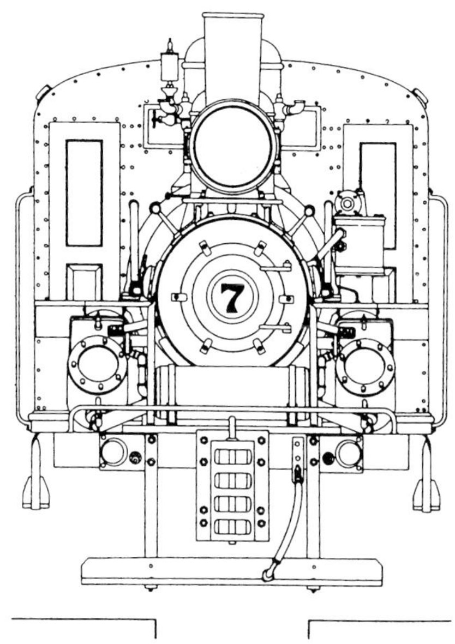

THE DOLLY VARDEN ORE CARS

On April 26, 1920, the Taylor Company placed an order for 20 ore cars with the Pacific Car and Foundry Company in Renton, Washington. These ore cars were to be wooden hopper cars, each with a 40,000-pound capacity. Chief Engineer Sawford who had formerly worked for the Canadian Collieries Railway based his design for these cars on the all-wood coal hopper cars used by Canadian Collieries (Dunsmuir), Limited, on their standard gauge Canadian Collieries line at Cumberland on Vancouver Island. A Taylor Engineering Company requirements drawing for these cars (now in the collection of John E. Lewis) bears a Canadian Collieries approval stamp and is dated February 3, 1920. Apparently Sawford used the Canadian Collieries plans outright without altering them. When they were completed in May of 1920, the cars were painted black with white lettering and numbered 100 through 119.

These wooden cars were so durable that the remains of some of them were still lying in the grass along the foreshore of Alice Arm in 1979, more than 60 years after they were built. Several more were buried in a slide at the mile 4 ballast pit and may be there yet.

With the arrival of the new wooden hopper cars designed by Sawford, the Dolly Varden Mines Railway began its banner shipping season of 1920. The railway shipped 28,700 tons of ore that year and could have gone on shipping that much (and more) each year if economic conditions had favored the continuation of mining.

Left, Top Right and Center Right:
These three photos show Dolly Varden Mines Railway wooden hopper car #116 sitting at the plant of her builder, the Pacific Car & Foundry Company in Renton, Washington. The car was 23 feet long, had a 40,000-pound capacity and weighed 20,300 pounds empty. Car #116 was built along with 19 other cars in 1920.

This interior view taken at the Pacific Car & Foundry plant in Renton, Washington, shows the door-operating mechanism of Dolly Varden Mines Railway wooden hopper car #116.

A string of Dolly Varden Mines Railway wooden hopper cars waiting at the Pacific Car & Foundry Company plant in Renton, Washington.

CERTIFICATION

The Dolly Varden Mines Railway finally received a provisional operating certificate from the Department of Railways on October 28, 1920. The line was officially opened under Certificate #404, issued on July 4, 1921. However, another inspection carried out by the Department on August 7, 1921, found a number of deficiencies: Locomotives 5, 6, and 7 were then in use and locomotives 8 and 9 were being repaired. Locomotive #5 was given a hydrostatic test of 215 pounds and her boiler was found to be tight, however, her steam gauge was several pounds off and had to be adjusted. Inspector Macintyre requested that the lower waterglass mounting be raised so its lowest reading would be 3 inches above the crown sheet to ensure an adequate safety margin in the boiler. He also required that a tender and automatic couplers be added to locomotive #5.

Dolly Varden Mines Railway
Wooden Hopper Car

Drawn by Rich Dunn

Scale: ¼ inch = 1 foot

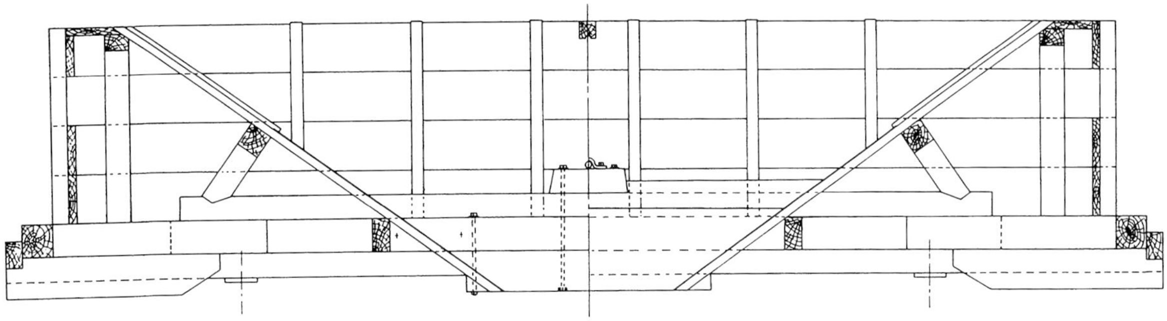

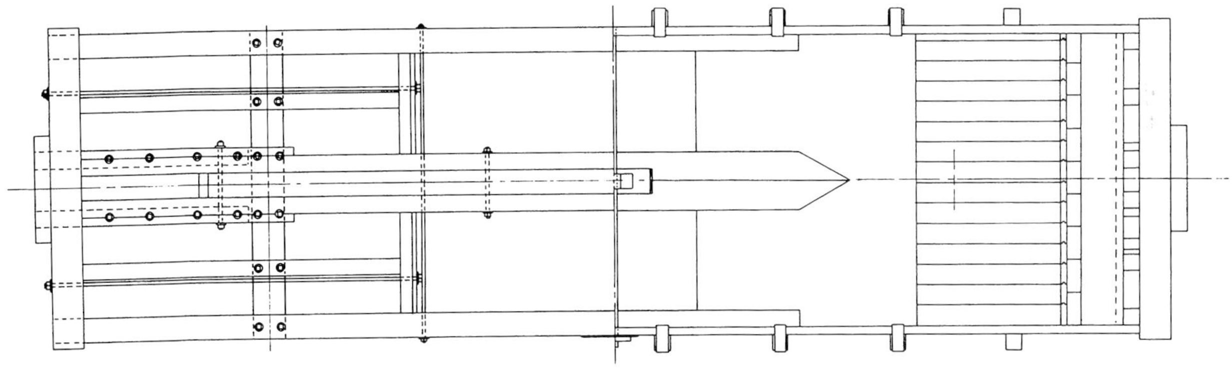

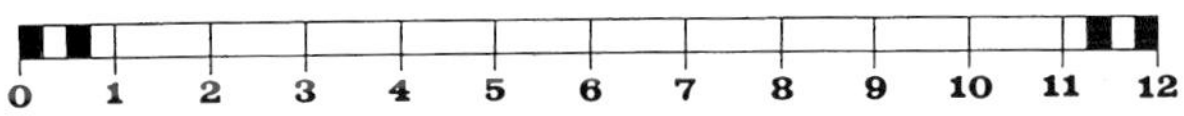

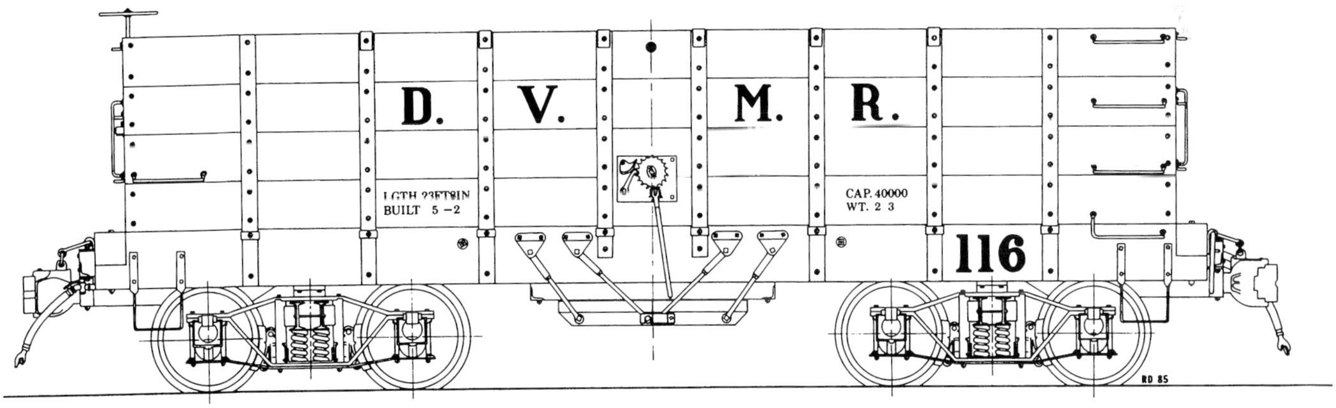

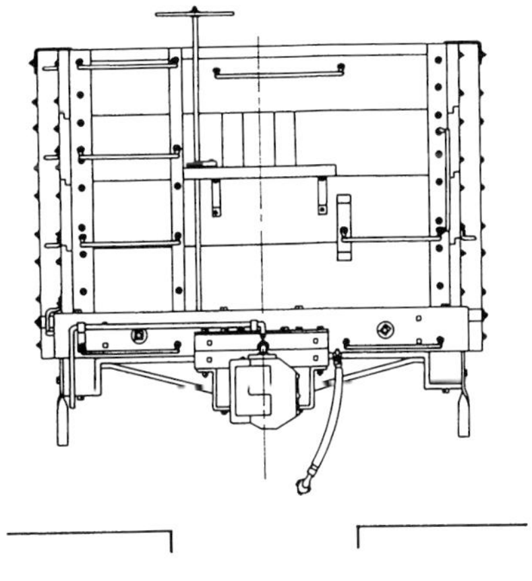

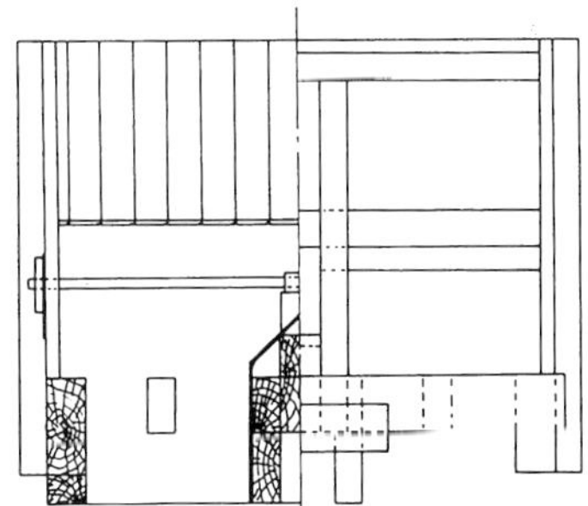

Number 116 was one of 20 cars ordered by the Dolly Varden Mines Railway in 1920 from the Pacific Car and Foundry Company of Renton, Washington.

Locomotive #6 was in a similiar condition to #5. She passed a 215-pound hydrostatic test but her try cocks and water-glass mountings were partially clogged and required cleaning. The water-glass mounting also had to be raised by 3 inches. Locomotive #6's blowoff cocks required overhauling and a handle so they could be operated. The washout plugs also had to be renewed. The boiler contained considerable scale and had not been cleaned out since the previous December. The tires and the rest of the machinery were in fair condition. The Inspector also required that a tender and automatic couplers be added to locomotive #6.

Locomotive #7 was given a hydrostatic test of 250 pounds and had a slight leak from the wrapper seams, around the door ring, and from a few of the boiler tubes. The seams were caulked and the tubes expanded to seal the leaks. The steam gauge was 5 pounds high and was adjusted. The stay bolts were okay. The tires and machinery were in fair condition but the crown and pinion gears were badly worn and required adjustment. Locomotive #7 also needed automatic couplers.

The wooden hopper cars were in good condition and were standing up well to rugged use. The flatcars needed automatic couplers and their air-brake equipment was in need of cleaning and overhaul. The boxcar/caboose also needed automatic couplers and air brakes. A conductor's valve was put inside the car so the brakes could be controlled by a trainman.

The items noted as requiring attention in this inspection were of the kind commonly neglected on industrial railways. While the Inspector found much that needed attention, this was not unusual. After the recommended changes were taken care of a working certificate was obtained. However, the railway operated for only a short period after this: the ore cars were not used after 1921 and the other equipment was used only during the few years when the Granby logging operations were being carried out at Alice Arm.

A ROSTER OF LOCOMOTIVES THAT RAN ON THE RIGHT-OF-WAY OF THE DOLLY VARDEN MINES RAILWAY

Road #	Type	Cyl. Size	Weight (tons)	Builder	Const. No.	Date	Tractive Effort (lbs.)	Prev. Owner
DVMR #1	0-4-0T	11 x 16"	12	Porter	?	?	?	Grand Trunk Pacific
DVMR #2	0-4-0T	11 x 16"	12	Porter	?	?	?	Grand Trunk Pacific
DVMR #3	0-4-0T	9 x 14"	14	Davenport	79	1903	6222	Granby (leased)
DVMR #4	(?) There is no record of a locomotive with this number.							
DVMR #5	0-4-0T	10 x 16"	21	Porter	4996	1911	8000 (?)	
DVMR #6	0-4-0T	10 x 16"	21	Porter	4997	1911	8000 (?)	
DVMR #7	30-2B	30		Climax	1543	1919	13,333	None
DVMR #8	2-4-2T	18		Baldwin	9763	1888	9000	Ferries & Cliff House
DVMR #9	2-6-0	?		Brooks	?	?	?	Utah Northern (?)
ALMD #3*	0-4-0T	?		Canadian Locomotive Co.	1218	1914	6667	J.A. Dewar
HM&D**	0-4-0	n.a.	4¼	Plymouth (gas/mechanical) (model FL2)	2320	1926	2000	(New)

* Abbotsford Lumber, Mining & Development Company
** Homestake Mining & Development Company

Scale: ¼ inch = 1 foot

This plan was traced by the author from a 1919 Taylor Engineering Company drawing of one of the Dolly Varden Mines Railway "Flatcars" that were actually flatcar/gondolas. The other two of the 15 DVMR Flatcars were only 10 feet long and 6 feet wide and no plans for them have been found.

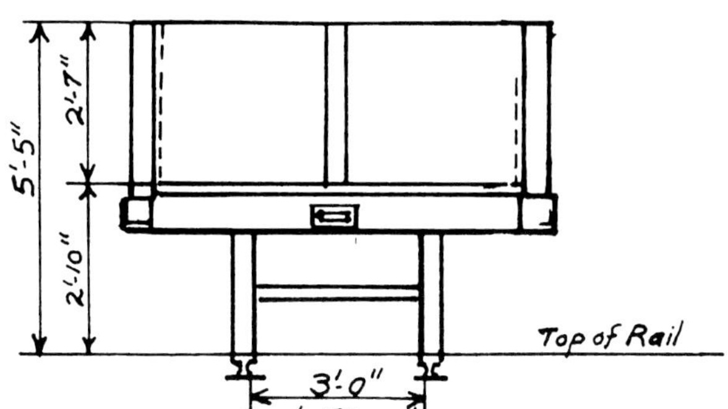

End Elevation

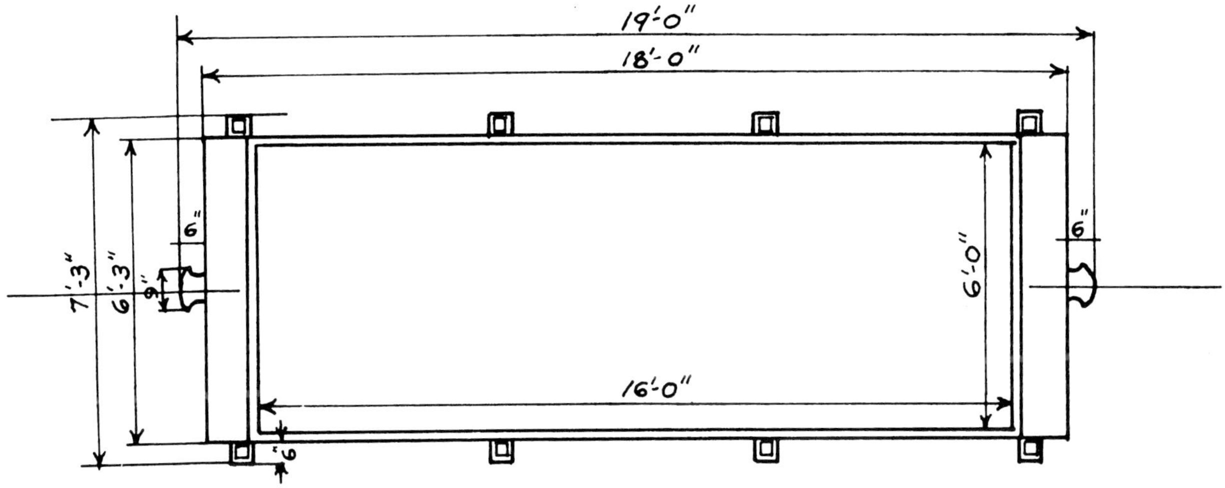

Plan.

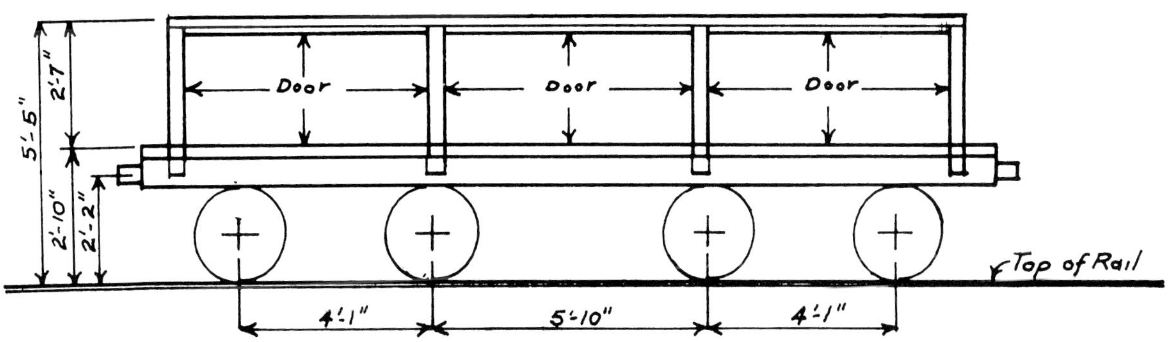

Side Elevation

Makers Name
U.S. Iron Works
Seattle U.S.A.
T. 10,000
CAP. 30,000

HISTORY OF THE DOLLY VARDEN MINES RAILWAY LOCOMOTIVES

The roster of locomotives gives many statistics for the locomotives that ran on the Dolly Varden Mines Railway right-of-way. The Abbotsford Lumber, Mining and Development Company shared trackage with the Granby logging operation on the Kitsault Flats during the logging years, 1923 to 1927. While ALMD #3 did not actually run on the Dolly Varden Mines Railway right-of-way *per se*, it is included because it ran just across the Kitsault River on the logging line. DVMR locomotives numbers 6 and 7 also ran on this logging line. The Consolidated Homestake Mining and Development Company used their 4½ ton, gas mechanical locomotive to pull trains of small 4-wheeled flatcars over the then provincially-owned rails of the former Dolly Varden Mines Railway well into the 1930s. Supplies arriving at Camp 8 on these trains were then packed up to the Homestake Mine near the Kitsault Glacier.

Dolly Varden Mines Railway Locomotives #1 and 2 are thought to have come from the railway contractor Foley, Welch and Stewart as surplus engines after the Grand Trunk Pacific Railway was completed to Prince Rupert in 1914. They were disposed of before the first official Dolly Varden Mines Railway inspection report in October of 1918. DVMR Locomotive #3 was leased from the Granby Company and was returned to its owners in September of 1919. DVMR numbers 5 and 6 were built to Canadian boiler standards for the C. W. Werdenhoff Company, an equipment dealer of Minneapolis, Minnesota. They were purchased secondhand by the Taylor Engineering Company in 1918 from an unknown previous owner. Both were scrapped at Alice Arm in 1937. DVMR Locomotive #7, a Climax, was the only locomotive built new for the Dolly Varden line. She was cut up for scrap at Alice Arm in 1937.

MILES OF RIGHT-OF-WAY AND NUMBERS OF DIFFERENT KINDS OF EQUIPMENT ON THE DOLLY VARDEN MINES RAILWAY IN RELATION TO TONS OF ORE SHIPPED 1917 – 1923

Year	Miles of Right-of-Way	Ballast (Dump cars)	Boxcars	Cabooses	Coal cars	Flatcars	Ore (Hopper) Cars	Steam Shovels	Locomotives	Tons of Ore Shipped
1917	8½	4	1	2	–	4	–	1	2	0
1918	15	4	1	2	–	4	–	1	3	0
1919	18	20	1	2	2	15	–	1	3	6,709
1920	18	20	1	–	2	15	20	1	5	28,700
1921-23	17½	20	1	–	3	15	20	1	5	1,874

The steam shovel and one of the locomotives (DVMR #3) were leased. The rest of the equipment was owned by the railway. The right-of-way decreased from 18 miles in 1920 to 17½ in 1921 because of the half-mile beyond the Dolly Varden Mine to the Wolf Mine that was being worked on in 1920 and was abandoned in 1921.

Dolly Varden Mines Railway numbers 8 and 9 were purchased from the Clayburn Company of Abbotsford, B.C. in February of 1920. They were rebuilt at the Vancouver Engineering Works before shipment to Alice Arm in March of 1920. When she was purchased DVMR #8 already had a long history of resales. Originally Ferries & Cliff House #8 (2nd.) of San Francisco, she was sold to the United Railways of San Francisco, who sold her to N.B. Livermore (a dealer), who sold her to the Vancouver contracting firm of Armstrong and Morrison in August of 1906. She was finally purchased by the Clayburn Company after December of 1906 when she was re-boilered by the Albion Machine Works at Vancouver, B. C. DVMR #8 worked on the

Clayburn mining railway until sometime between 1915 and 1920 when she and her stablemate were retired in favor of electric motive power. Locomotive #9 was a Brooks Mogul purchased by the Armstrong and Morrison Company about 1906. Little is known about her early history but she may have come from the Utah Northern which had a number of surplus Brooks Moguls for sale about that time. Like #8, she was re-boilered by the Albion Machine Works before being purchased by the Clayburn Company. DVMR numbers 8 and 9 were scrapped at Alice Arm in 1937.

Abbotsford Lumber, Mining and Development Company Locomotive #3 was purchased used from the B.C. Equipment Company. She had been constructed in 1914 for the J. A. Dewar Company of Toronto as their #3. The final disposition of ALMD #3 is unknown, but she is known to have been removed from the Kitsault Flats in 1928 after sitting idle for some time.

The Consolidated Homestake Mining and Development Company gas/mechanical locomotive was built new for them. Her fate is not known but she pulled supply trains up to Camp 8 well into the 1930s.

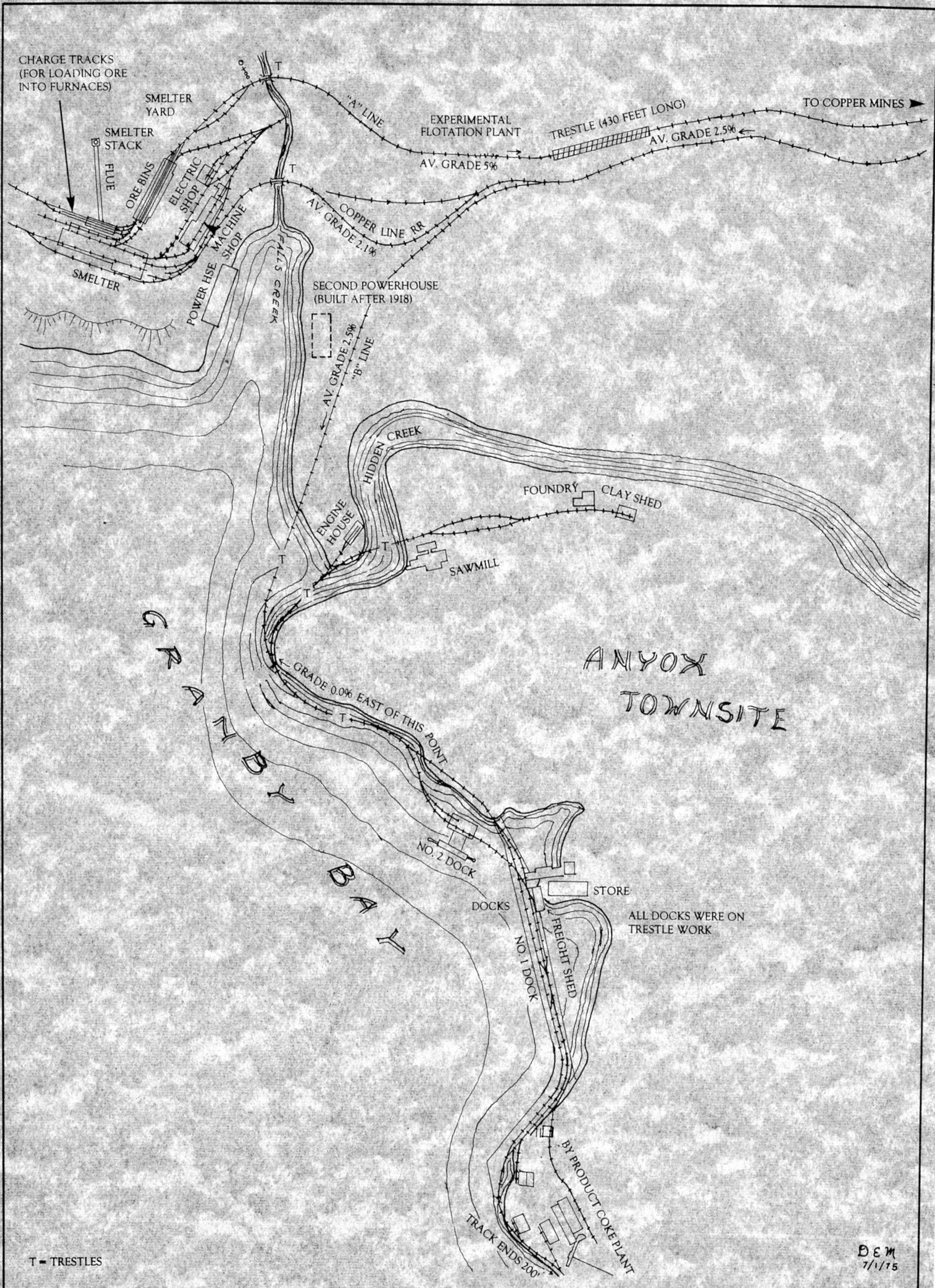
CHARGE TRACKS (FOR LOADING ORE INTO FURNACES)
SMELTER YARD
SMELTER STACK
FLUE
ORE BINS
ELECTRIC SHOP
MACHINE SHOP
SMELTER
POWER HSE.
"A" LINE
EXPERIMENTAL FLOTATION PLANT
TRESTLE (430 FEET LONG)
TO COPPER MINES
AV. GRADE 5%
AV. GRADE 2.5%
COPPER LINE RR
AV. GRADE 2.1%
FALLS CREEK
SECOND POWERHOUSE (BUILT AFTER 1918)
AV. GRADE 2.5%
"B" LINE
HIDDEN CREEK
FOUNDRY
CLAY SHED
ENGINE HOUSE
SAWMILL
ANYOX TOWNSITE
GRANBY BAY
GRADE 0.0% EAST OF THIS POINT
NO. 2 DOCK
DOCKS
STORE
ALL DOCKS WERE ON TRESTLE WORK
NO. 1 DOCK
FREIGHT SHED
BY PRODUCT COKE PLANT
TRACK ENDS 200'
T = TRESTLES
DEM
7/1/75

B

Appendix B: Notes on the Railway Operations and Equipment of the Granby Consolidated Mining, Smelting and Power Company at Anyox, B.C.

by John E. Lewis

Much of the ore taken from the Dolly Varden Mine was processed at the mining and reduction works of the Granby Consolidated Mining, Smelting and Power Company located at Anyox, on Observatory Inlet about 18 miles from Alice Arm. The Granby Company played several roles in the Dolly Varden story: as a potential purchaser of the mine and railway, as a lessor of equipment, and as one of the companies processing the ore. Therefore it is fitting to include some information on this company whose fortunes were so closely entwined with those of the Dolly Varden Mine.

HISTORY OF THE ANYOX OPERATIONS

Rich copper deposits at Anyox were first prospected at the turn of the century and the first claims were staked in 1901 by a group of men from Port Simpson, just north of Prince Rupert. The three claimants, Messrs. McMillan, Rudge, and Flewin, called themselves the "Hidden Creek Group" after the name Anyox, which means "Hidden Creek" in the Nishga Indian language. Two creeks – Hidden Creek and Falls Creek – flow into Granby Bay at the Anyox town site.

In 1902 these properties were "bonded" for $40,000.00 to M.K. Rodgers of Seattle, an agent for the estate of Marcus Daly of Montana mining fame. Daley had risen from a pick-and-shovel man to become one of the richest mine owners in western North America. After he died in 1900 his heirs were not able to continue his company's pace of mine acquisition and development. Although some development was made under this lease, the properties soon reverted back to the Hidden Creek Group.

In 1905 a syndicate was formed in Vancouver that "bonded" the properties for a fee of $25,000.00. Three years later, M.K. Rodgers purchased the properties again, this time in partnership with Thomas Hodgins, a Butte, Montana banker. The price had then risen to $135,000.00.

Finally, in 1910 the properties were bought by the Granby Company for $500,000.00 and large-scale development work began. A townsite, mines, railway, and coking and smelting facilities were established. Production began in 1914 and by the 1920s the town of Anyox had a population of 3000. The smelter soon became the largest copper smelter in the British Empire. The prosperity of Anyox continued until 1935 when low copper prices during the Great Depression caused the mines and smelter to close. The abandoned remains of the mines and smelter were destroyed in a forest fire in 1942.

Left:
A portion of the Granby narrow gauge railway at Anyox, B.C. around 1918. The "A" Line ran from the Mines to the smelter and the "B" Line from the "A" Line to the Wharf. The gauge was 36 inches.

A winter scene at the Granby Consolidated Mining, Smelting and Power Company concentrating mill at Anyox, B. C. Note the dead trees killed by toxic emissions from the smelter.

THE GRANBY NARROW GAUGE RAILWAY AT ANYOX

The Granby mines and smelter were connected by an extensive 36-inch narrow gauge railway that had 3.22 miles of mainline and 2.36 miles of sidings and yard trackage. The track layout was as follows:

The mainline track –

The 7913-foot-long "A" line had a switchback and connected the mines and smelter.

The 7490-foot-long "B" line had an average 2.1 percent grade and connected the docks with the "A" line.

The 1516-foot-long "Copper Line" connected the smelter and the "B" line.

The sidings and yard track –

The smelter yard trackage was 2626 feet long.

The ore and coke tracks on the high line were 2753 feet long.

The charge tracks were 3167 feet long.

The sawmill and foundry spur was 1770 feet long.

The dock tracks were 1710 feet long.

There were also 458 feet of storage sidings.

There were 1.95 miles of trestle work on the line and they were laid with 56-pound rail.

The equipment initially consisted of two 42-ton Baldwin Westinghouse electric locomotives, 25 steel ore cars and some miscellaneous equipment. By 1920 this

One of three Granby Baldwin/Westinghouse 36-inch gauge electric locomotives used at Anyox. Hidden Creek Copper Company #21, a B-B type, was built in 1913 and bore the construction number 39986. She weighed in at 84,000 pounds, was rated at 300 h.p. and developed 18,666 pounds of tractive effort. Hidden Creek Copper Company, a subsidiary of Granby, was named for Anyox, which means "Hidden Creek" in the Nishga Indian language.

Granby electric locomotive #10 (a Baldwin/ Westinghouse, 36-inch gauge) entering the Hidden Creek Mine.

Granby steam locomotive #5 at Anyox. She was an Alco 0-6-0.

This little Granby 0-4-0T (unidentified as to number) was photographed at Anyox.

equipment had been supplemented with new equipment and locomotives from another Granby operation, their mines and smelter at Phoenix in the Grand Forks area of British Columbia north of Spokane near the Washington State border. (Phoenix was 20 miles northwest of Grand Forks and has been abandoned.)

The 1920 report of the Provincial Department of Railways listed 6 miles of track, eight steam locomotives and three electric locomotives. Also listed were two boxcars, 50 ore cars, three coal cars, 20 ballast cars and a steam shovel. Granby steam locomotive #3 was leased to the Dolly Varden Mines Railway during its 1918-1919 construction period and may also have been used on the Granby narrow gauge logging operations on the Kitsault Flats at Alice Arm.

Much of the Granby Company's rolling stock at Anyox and on its logging operation at Alice Arm was constructed by the Seattle Car and Foundry Company and its successor the Pacific Car and Foundry Company of Renton, Washington. On December 31, 1914, the Granby Company ordered three 34-foot, 50,000-pound capacity flatcars. The cars were built to narrow gauge standards of size and safety rather than to industrial line standards. Each car was equipped with Sharon-type, bottom-operating, automatic couplers and Westinghouse automatic air brakes.

A 36-inch gauge Granby 34-foot steel flatcar built by the Seattle Car and Foundry Company in December of 1914.

A 36-inch gauge Granby Company 26-foot flatcar built by the Seattle Car and Foundry Company in December of 1914.

A 36-inch gauge ore car built by Kilbourne and Jacobs of Columbus, Ohio, in March of 1915 and supplied to Pacific Car and Foundry Company for delivery to the Granby Company as part of their first order for ore cars.

Granby's first order for ore cars was placed with the Seattle Car and Foundry Company on March 16, 1915. The design for these cars apparently came from the Kilbourne and Jacobs Company of Columbus, Ohio. This order for 10 cars was the first in a series of orders that eventually produced 25 ore cars of this design for the Granby Company's Anyox operation. All of these Granby ore cars were of all-steel construction and carried 370 cubic feet of ore with a maximum load of 80,000 pounds. At least two and perhaps all of these cars were built not by the Seattle Car and Foundry Company but by Kilbourne and Jacobs. The Seattle company had photos of the cars but most if not all of the construction was sub-contracted to the Ohio firm. The road numbers assigned to these cars are not known at this writing.

Four 26-foot flatcars were ordered on July 20, 1915, that were identical to the first two flatcars except for their shorter length. They were numbered 301 to 304.

An order for 15 steel ore cars was placed on January 24, 1916, with the Seattle Car and Foundry Company. These cars were identical to the first steel ore cars built for the Granby Company and apparently were actually built by the Kilbourne and Jacobs Company. The numbering of these cars is unknown; however, it was probably in the 200s since one car in the series bore the number 237.

Granby placed no more orders with the Seattle Car and Foundry Company until after it had been renamed the Pacific Car and Foundry Company in mid-1917. In April

This broadside view of one of the 36-inch gauge Granby ore cars built by Kilbourne and Jacobs in 1915 shows the distinctive K & J builder's plates.

This 36-inch gauge Granby steel ore car #237 was one of 15 ordered from the Pacific Car and Foundry Company in January of 1916. Her hopper doors are open.

of 1923, four side-dump ore cars with a completely new design were ordered. These cars were designed to carry 16-cubic yards (432 cubic feet) weighing no more than 50,000 pounds. The last of these cars was delivered in October of 1923. These cars were assigned road numbers 501 through 504.

The final order from the Granby Company to the Pacific Car and Foundry Company was on January 23, 1924 and was for ten sets of 2-bunk connected trucks (to form ten 2-bunk log cars). The design of these cars was identical to that of two 2-bunk cars that Pacific Car and Foundry had built in 1923 for the firm of Brooks, Bidlake, Whithall, Limited of Lang Bay, B.C. (Lang Bay is 10 miles east of Powell River

This bottom view of one of the Granby ore cars built by Kilbourne and Jacobs shows the bolster and bolster pin, the closure mechanism and the hinges for the center-dump hopper doors.

A pair of the 36-inch gauge Granby ore cars built by the Pacific Car and Foundry Company from a design of the Kilbourne and Jacobs Company of Columbus, Ohio. The running boards were on the left side of each car.

and 75 miles north of Vancouver. Lang Bay was the site of a narrow gauge logging railway until the late 1920s.) It was also identical to the design of some cars built by the Swayne Lumber Company of Oroville, California in 1917 and 1919 from 65 sets of steel parts furnished by the Pacific Car and Foundry Company.

The Pacific Car and Foundry Company provided the Granby Company with the trucks, couplers, bunks, air-brake equipment and miscellaneous hardware for these log cars. The wooden parts were made by the Granby Company which also assembled the cars. The length of these cars is not known but, on the basis of standard Pacific Car and Foundry designs, we would expect them to have been at least 38 feet long – the same length as the Swayne Lumber Company log cars. The numbering of these cars is unknown; they were used by the Granby Company on their logging operations on the Dolly Varden Mines Railway and on the Kitsault Flats.

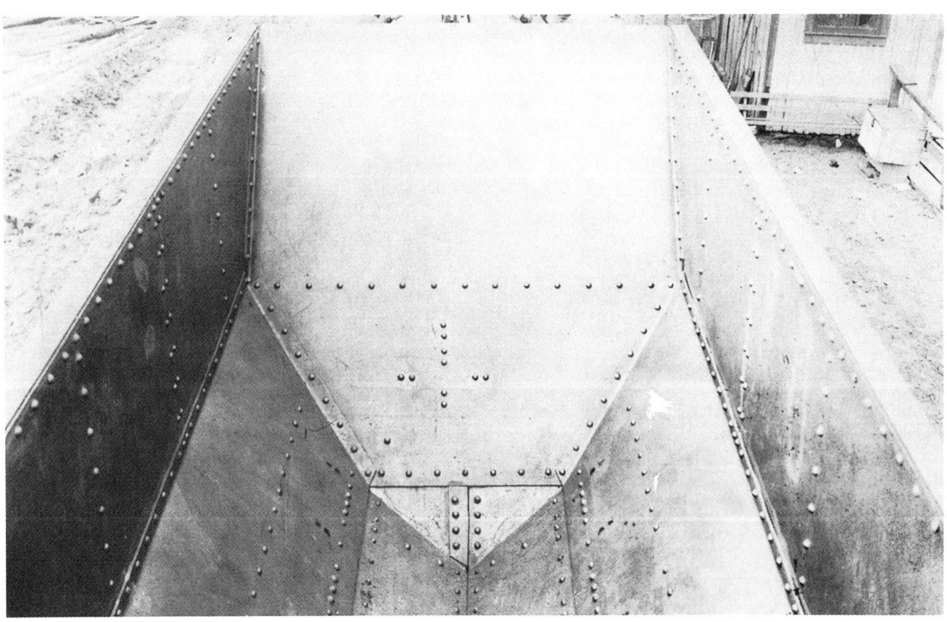

The inside of a 36-inch gauge Granby ore car of Kilbourne and Jacobs design, showing the hopper interior and the rugged construction of these durable cars.

GRANBY CONSOLIDATED
Mining, Smelting and Power Company
Steel Ore Car

Built by the
Pacific Car and Foundry Company
of Renton, Washington in 1915.

Drawn by Ron Anstey

Scale: ¼ inch = 1 foot

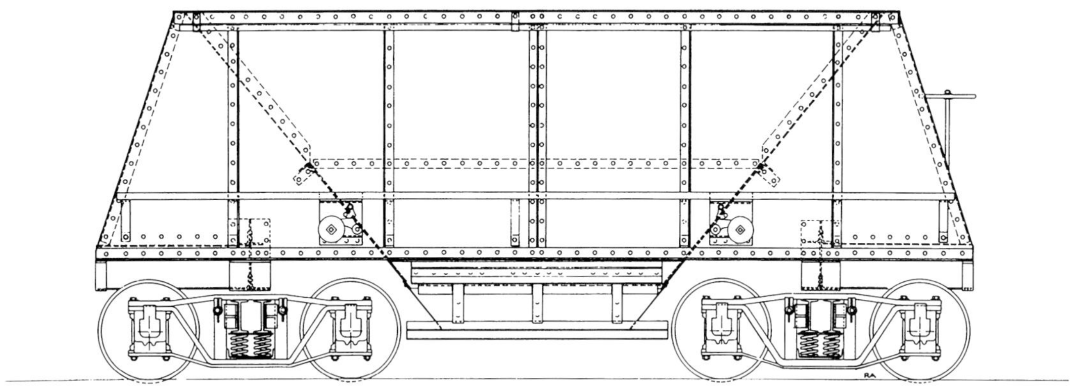

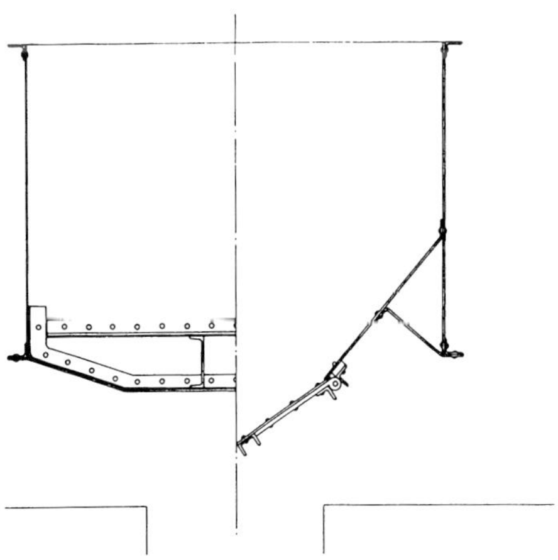

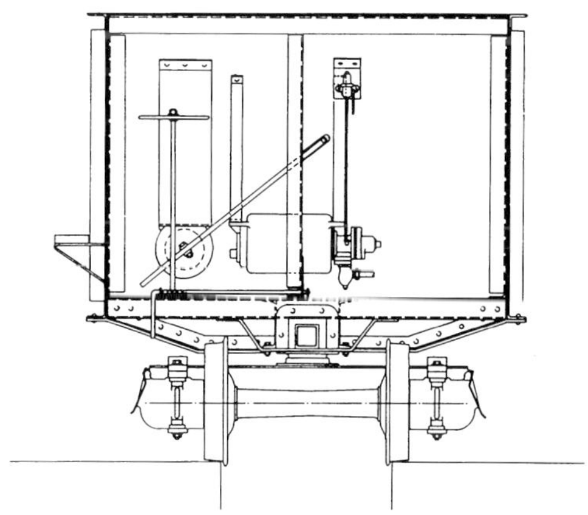

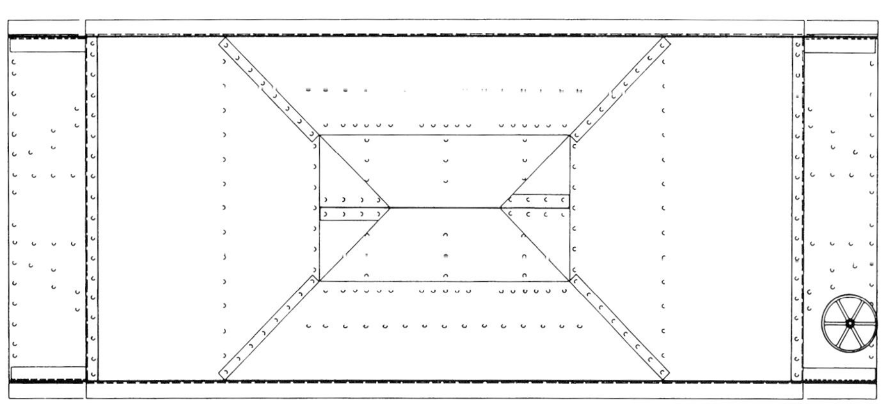

One of the four 36-inch gauge side dump ore cars built in 1923 for the Granby Company by the Pacific Car and Foundry Company.

OPERATIONS ON THE GRANBY NARROW GAUGE RAILWAY

The Granby narrow gauge railway operated both steam and electric locomotives. It was well designed and had an impressive amount of trestle work. Besides two major trestles on the "A" line, there were three major trestles at the mouth of Hidden Creek, and the wharf trackage was almost all on trestles. The company maintained high standards of construction: no curves greater than 28 degrees were permitted. The minimum width of cuts was 12 feet for earth cuts and 10 feet for rock cuts. Even though the track gauge was only 36 inches, the subgrade for the track was at least 8 feet wide and in many places 10 feet wide. The overhead wires for electric operations were supported by cedar poles placed 55 feet apart. The poles were offset 7 feet, 6 inches from the centerline of the track; mast arms held the wire over the center of the track.

Apparently the operation of the Granby narrow gauge railroad was as well engineered as was its construction. In 1918 the Minister of Mines for British Columbia reported that the Hidden Creek Mine was producing 65,000 tons of ore per month for the smelter at Anyox. At 25 tons per ore car this would have required 2600 carloads per month; that's 260 ten-car loaded trains, and an equal number of empty trains returning. Assuming a 6-day week, that's about 25 trains per day or one every 20 minutes or so of an 8 hour work day. And that is only the operation on the "A" line.

The "B" line ran between the wharf and the smelter. It hauled in general supplies, timber, ore from other mines, coke and limestone needed for smelting, and hauled out the finished product: blister copper. Each month of 1916, at least 266 carloads of general merchandise and mining equipment were hauled from the wharf to the town and to the various Granby operations. Timber from the Granby logging operations (including those at Alice Arm and on the Kitsault Flats) was hauled from the wharf to

The experimental flotation plant at Anyox. A small 0-4-0 saddle-tank locomotive is starting across the trestle on the right. This 430-foot long trestle was on the "A" line.

This map of the Granby Smelter Plant and the Anyox area was copied from the *Granby News* Vol. 1, #3, 1917.

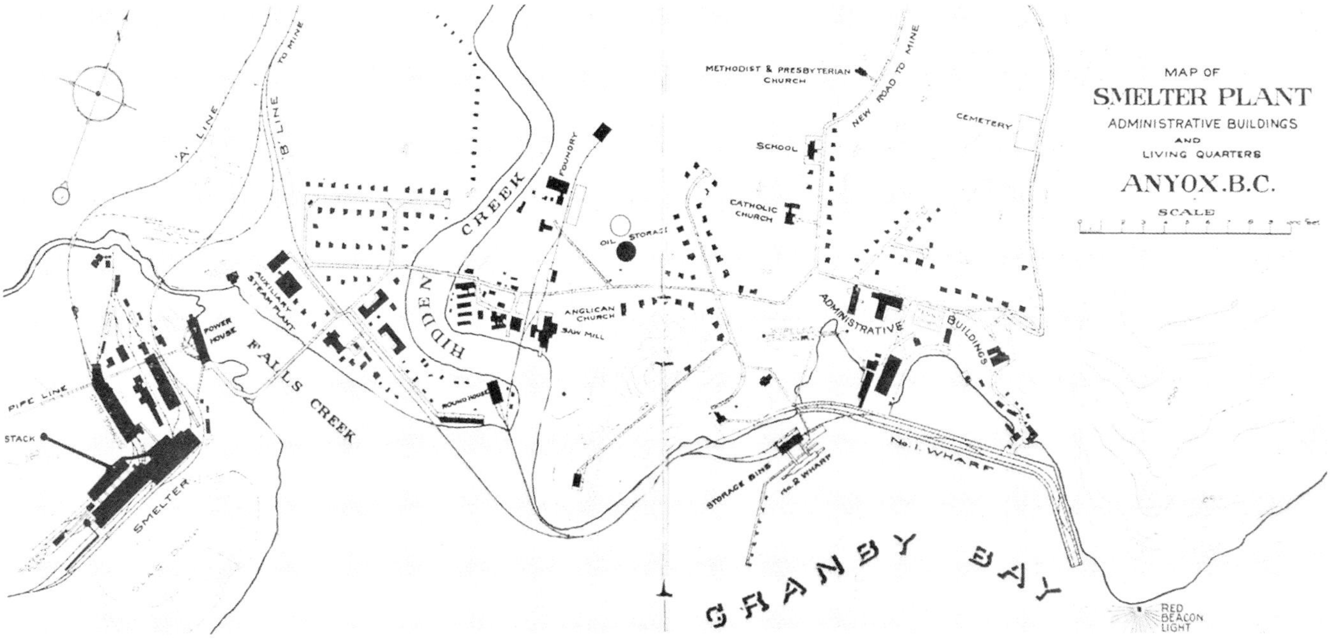

A 1918 view of the Anyox smelter. The main powerhouse is just below the smelter and the flume on the right foreground carries water to powerhouse #2. The "Copper Line" crosses Falls Creek above the flume on the right.

The Granby Company docks at Anyox in 1919.

The Granby Company smelter site at Anyox.

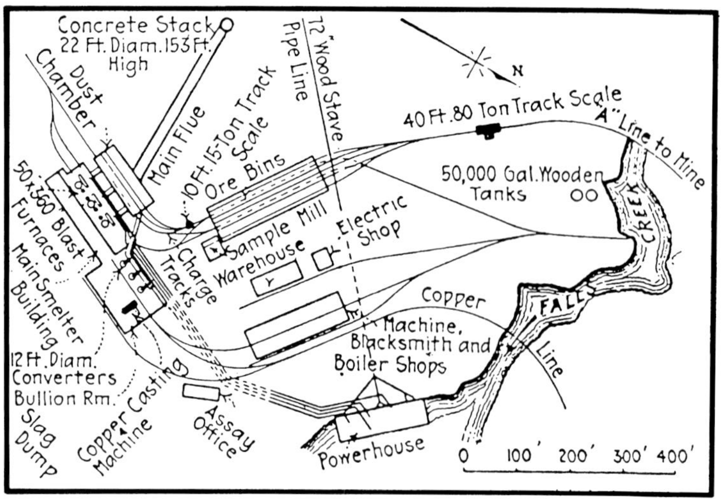

This 1917 view shows one of the two Granby docks at Anyox; some of the company's administration buildings are on the right.

The harbor at Anyox in 1917.

the sawmill, and mine props and construction lumber was hauled from the mill to sites where they were needed. At peak production the daily output of the sawmill was around 20,000 board feet, about nine carloads of lumber per day. In 1919 at least 400 carloads of ore were received each month from the Dolly Varden Mine and from other mines in British Columbia and southeast Alaska. To process 75,000 tons of ore per month the smelter would have required at least 158 carloads of coke and 178 carloads of limestone. About 14 carloads of copper were produced each month in 1914 and 1915. Including returning empties, all this adds up to over 2000 cars per month being hauled on the "B" line.

This intensive activity meant that train movements had to be carefully controlled to avoid collisions. Each train was assigned a conductor and there was an efficient electric block signal system. At the time, this was the only electrical block signal system in use on a narrow gauge railroad in either Canada or the U.S.A. The *Granby News* (February, 1918) reported that Granby train crews used this system "with ease and success, minimizing accidents to a remarkable degree."

The Characters by Affiliation

The Dolly Varden Mines Company of Chicago

Anonymous Investors (Chap. 2)
Mr. Brewster, a member of the stockholder's committee (Chap. 5)
R. T. Elliott, Counsel (Chap. 10)
Mr. Hamill, Chicago attorney (Chaps. 5, 7, 8)
Herbert Hoover, thirty-first president of the U.S.A., Investor (Chap. 2)
Joseph Hubbard, President (Chaps. 2, 3, 4, 5, 6, 7, 8, 13)
C. C. Lindberg, Consulting Engineer at the Dolly Varden site (Chaps. 2, 4)
R. B. (Richard) McGinnis, Resident Engineer at the Dolly Varden site (Chaps. 2, 3, 4, 5, 6, 8) – see also Taylor Mining Company and Miscellaneous Affiliations
L. W. Patmore, Vancouver solicitor (Chaps. 2, 3, 4, 8)
Joseph Wingfield, Director and Investor (Chap. 2) – see also Goldfield Consolidated Mines Company

Officials of the Government of the Province of British Columbia

A. L. Carruthers, District Engineer of Public Works for the Southern District of Atlin (Chap. 10)
J. S. Cowper, member, select committee of the Provincial Legislature (Chap. 8)
J. W. de B. Farris, Minister of Labor (Chap. 7)
D. G. Forbes, Department of Mines (Chap. 1)
Dr. J. H. King, Minister of Public Works (Chap. 9)
A. McPhee, Inspector (Chap. 10)
R. W. Macintyre, Inspector of Railways (Chaps. 7, 8, App. A)
W. A. MacKenzie, member, select committee of the Provincial Legislature (Chap. 8)
Mr. Mobley, member of the legislature (Chap. 10)
Justice Morrisson of the Supreme Court of British Columbia (Chap. 10)
John Oliver, Premier and Minister of Railways (Chaps. 4, 7, 8, 10, 11)
F. A. Pauline, member, select committee of the Provincial Legislature (Chap. 8)
A. F. Proctor, Chief Engineer of Railways (Chaps. 4, 5, 6, 7, 9, 10, App. A)
William Rae, Chief Inspector of Equipment for the Department of Railways (Chaps. 9, 10)
David Whiteside, member, select committee of the Provincial Legislature (Chap. 8)
John Yorston, member, select committee of the Provincial Legislature (Chap. 8)

The Taylor Engineering Company of Vancouver

Anonymous Workers (Chap. 4)
Jasper S. Connell, Vice President and Construction Manager of the Dolly Varden Mines Railway (Chap. 6)
William McLean, resident engineer at Anyox (Chap. 2), Chief Engineer (Chaps. 3, 4)
Simon Maloney, Construction Foreman of the Dolly Varden Mines Railway (Chap. 6, 7)
E. C. Mayers, Solicitor (Chap. 8)
John Seymour, Assistant Engineer (Chap. 4)
Alfred J. T. Taylor, Founder and President (Chaps. 2, 3, 4, 5, 6, 7, 8, 13) – see also Taylor Mining Company
E. G. Taylor, brother and partner of Alfred Taylor (Chap. 2, 11)

The Granby Consolidated Mining, Smelting and Power Company

Mr. Campbell, Engineer (Chap. 6)
Mr. Earl, Chief Executive, New York office (Chap. 7)
F. L. Housely, operator of the logging near Alice Arm from May of 1926 to August of 1927. (Chap. 12)
Don Klanahan, director of logging operations on the Kitsault Flats and on the west side of the Dolly Varden Mines Railway in 1924. (Chap. 12)
D. McQueen, high-rigger for the lumber operation on the Kitsault Flats (Chap. 12)
O. B. Smith, Manager of Mines (Chap. 6)
F. M. Sylvester, Vice President (Chaps. 5, 6, 7, 8)
Mr. Todd, Assistant to O. B. Williams (Chap. 6)
O. B. Williams, Superintendent of Smelters (Chap. 6)
Mr. Wooster, Treasurer (Chap. 6)

The Goldfield Consolidated Mines Company of Reno, Nevada

George Wingfield, President (Chaps. 5, 8, 10, 11) – see also Dolly Varden Mines Company
Joseph Hubbard, Vice President (Chap. 5) – see also Dolly Varden Mines Company

The Taylor Mining Company

John Anderson, Secretary (Chap.11)
R. P. Butchard, Director, also president of the B.C. Cement Company (Chap. 9)
Hall C. Chiene, Director and trustee for the company's creditors (Chaps. 9, 10) – see also Miscellaneous Affiliations
Tom Dalrymple, Construction Engineer (Chap. 9)
Maj. Angus Davis, Superintendent of Mines (Chap. 9, 10)
Conductor Dunn (Chap. 10)
Engineer McCauley (Chap. 10)
R. B. (Richard) McGinnis, consulting engineer, Dolly Varden Mines site (Chap. 9) – see also Dolly Varden Mines Company and Miscellaneous Affiliations
Maj. C. B. North, mining engineer (Chap. 10)
C. M. Ralston, trustee for the company's bondholders (Chap. 10)
C. H. Rowlson, Director, also general manager of the Imperial Oil Company (Chap. 9)
Frank Sawford, Chief Mechanical Engineer (Chaps. 9, 10, App. A)
Lt. J. W. Smith, Assistant Business Manager (Chap. 9)
Alfred Taylor, President (Chap. 9, 10, 11) – see also Taylor Engineering Company
J. F. Tener, Director, also general manager of the Pacific Lime Company (Chap. 9)
Maj. D. Desmond Young, Business Manager (Chap. 9)

Explorers

The members of the Lyall Expedition (Chap. 1)
George Vancouver (Chap. 1)

Missionaries

Archdeacon Collinson (Chap. 1)
Alice Tomlinson (Chap. 1)
Robert Tomlinson (Chap. 1)

Miners, Prospectors, and Other Workers

Gus Anderson, miner (Chap. 1)
Jack Anderson, prospector (Chap. 1)
Donald Cameron, the prospector who sold the Wolf Group of claims to the Dolly Varden Mines Company in 1916 (Chap. 2)
E. Carlson, one of the four Scandinavian prospectors who discovered the Dolly Varden Mine (Chap. 1)
W. J. Crawford, miner (Chap. 1)
William Dilworth, prospector (Chap. 1)
Ludwig Eik, one of the four Scandinavian prospectors who discovered the Dolly Varden Mine (Chap. 1)
Ole Evindsen, one of the four Scandinavian prospectors who discovered the Dolly Varden Mine (Chaps. 1, 2) – see also Citizens of Alice Arm
Oscar Flint, miner (Chap. 1)
Mr. Frizzell, miner (Chap. 1)
Mrs. Handley, cook at the Dolly Varden Mine (Chap. 2)
Andrew Hanson, miner (Chap. 1)
Mr. Jacobsen, miner (Chap. 2)
Frank Jones, prospector (Chap. 1)
Fred Juggins, prospector (Chap. 1)
Jack Marshall, miner(Chap. 2)
Axel Norman, miner (Chap. 2)
Ole Pearson, one of the four Scandinavian prospectors who discovered the Dolly Varden Mine (Chaps. 1, 2)
Frank Roundy, prospector (Chap. 1)
George Salt, miner (Chap. 1)
John Stark, prospector and the first permanent resident of Alice Arm (Chap. 1)
John Strombeck, prospector (Chap. 11)
Charlie Swanson, prospector and miner, traded a grubstake for 1/5 interest in the Dolly Varden Mine (Chaps. 1, 2)
Joe Wells, prospector (Chap. 1)
Benny Williams, prospector (Chap. 1)
George Young, prospector (Chap. 1)

Miscellaneous Affiliations

J. Anderson, director of repairs on the Dolly Varden Railway in 1922 (Chap. 10)
Donald W. Cameron, Wolf Group of claims (Chap. 8)
Hall C. Chiene, accounting firm of Buttar & Chiene (Chap. 7) – see also Taylor Mining Company
A. C. DesBrisez, Attorney for George Wingfield for Northern Mining Properties, Limited (Chap.11)
Joseph Errington, Temiskaming Mining Company (Chap. 7)
A.C.H. Gerhardi, Manager For the Toric Mines Company (Chap. 11)
Thomas Hodgins, a Butte, Montana, miner, who with M.K. Rodgers, held the Anyox properties between 1908 and 1910. (App. B)
Mr. Hopkins (Chap. 1)
R. B. (Richard) McGinnis (Chap. 1) – see also Dolly Varden Mines Company and Taylor Mining Company
A. D. McPhee (Chap. 1)
H. W. Petri, Toronto equipment dealer (Chap. 2)
M. K. Rodgers, Marcus Daley estate (App. B)
Victor Spencer, Victor Spencer and Associates, holder of the Dolly Varden Properties from 1923 until his death in the 1950s (Chaps. 11, 13)
The Reverend Dr. D. W. Taylor, Alfred Taylor's father (Chap. 2)

Citizens of Alice Arm

Ole Evindsen, owner of the Alice Arm Hotel and foreman on the railway in the 1930s (Chap. 13) – see also Miners and Prospectors
Svea Evindsen, Ole Evindsen's wife (Chap. 2)
Al Falconer, owner of the general store (Chaps. 10)
T. W. Falconer, lessor of the Dolly Varden Mine, 1935 to 1940 (Chap. 13)
P. F. Kergin (Chap. 7)
Mrs. McKinley, early resident of Alice Arm (Chap. 1)
Mrs. Peterson, early resident of Alice Arm (Chap. 1)
Mrs. Wheatly, early resident of Alice Arm (Chap. 1)

The Consolidated Homestake Mining and Development Company

J. Strombeck, speeder passenger (Chap.11)
T. Tavener, speeder passenger (Chap.11)
F. Ness, speeder passenger (Chap.11)
J. Trinder, speeder passenger (Chap.11)

The Hidden Creek Group

Mr. McMillan (App. B)
Mr. Rudge (App. B)
Mr. Flewin (App. B)

Chronology

Preliminaries – 1793 to April 1917

1793	George Vancouver, explorer, is the first person of European descent to visit the Alice Arm area (Chap. 1)
1869	Alice Arm is named for the missionary, Alice Tomlinson (Chap. 1)
1910	The four Scandinavian prospectors discover the Dolly Varden Mine (Chap. 1)
1915	R.B. McGinnis leases the Dolly Varden Mine for $50,000.00 on behalf of a group of Chicago financiers (Chap. 1)
Fall of 1915	The Dolly Varden Mines Company is founded in Chicago and incorporated in Delaware (Chap. 2)
April 1916	William McLean, Engineer for the Taylor Company, estimates the cost to build the Dolly Varden Mines Railway at $113,535.00 (Chap. 2)
November 1916	Thousands of tons of ore have been "proven-up" at the Dolly Varden Mine (Chap. 2)
January 1917	At the request of Hubbard, the Taylor Company begins to order equipment for the Dolly Varden Railway (Chap. 2 and 3)
February 1917	Hubbard persuades Taylor to sign a contract in which the Taylor Engineering Company agrees to build the Dolly Varden Mines Railway for a sum not to exceed $175,000.00 (Chap. 3)
February 1917	Unwritten gentlemen's agreement between Taylor and Hubbard that the Dolly Varden Mines Company will pay expenses in excess of $175,000.00 that might be incurred in the construction of the Dolly Varden Mines Railway (Chap. 3)
April 1917	The act to grant a charter to the Dolly Varden Mines Railway receives its first reading, its final reading, and is written into law (Chap. 4)

Construction of the Dolly Varden Railway – April 1917 to July 1919

April 1917	John Seymour and William McLean arrive in Alice Arm with workers and begin to construct the Dolly Varden Mines Railway (Chap. 4)
July 1917	C.C. Lindberg prepares report showing Dolly Varden Mine site has proved-up over 50,000 tons of ore worth over one million dollars (Chap. 4)
September 1917	Cost of Dolly Varden Mines Railway exceeds the agreed upon amount of $175,000.00 by $31,000.00 (Chap. 4)
October 1917	First 8 miles of the Dolly Varden Mines Railway are completed (Chap. 4)
December 1917	The Taylor Engineering Company overdraws its bank account by $12,000.00 in building the Dolly Varden Mines Railway (Chap. 5)
January 1918	Alfred Taylor accepts a cash settlement from the Dolly Varden Mines Company on Hubbard's advice and because he needs the money (Chap. 5)
January 1918	The Dolly Varden Mines Company management decides not to proceed with the railway and to sell its interest in the Dolly Varden Mine and Railway (Chap. 5)
January 1918	Alfred Taylor asks the Granby Consolidated Mining, Smelting and Power Company to take over and complete the railway (Chap. 5)
May 1918	Taylor resumes construction of the Dolly Varden Mines Railway with $50,000.00 of Taylor Engineering Company money (Chap. 6)
July 1918	The Taylor Company spends 30 or 40 thousand dollars more than the $50,000.00 planned and borrows money from the Granby Company in order to proceed with the construction of the Dolly Varden Mines Railway (Chap. 6)
August 1918	Simon Maloney, Construction Foreman, widens the Dolly Varden Railway right-of-way from 8 to 10 feet at the presumed request of F.M. Sylvester, the Granby Company vice president (Chap. 6)
September 1918	Alfred Taylor rushes to Chicago to be on the spot when the Granby Company buys the Dolly Varden Mine and Railway only to find that the sale is off (Chap. 7)
October 1918	The Taylor Engineering Company declares bankruptcy (Chap. 7)
October 17, 1918	First inspection of the Dolly Varden Mines Railway by R.W. Macintyre for the government of British Columbia
October 22, 1918	Work ceases on the Dolly Varden Mines Railway leaving over 300 men out of work and unpaid (Chap. 7)
January 1919	The Temiskaming Mining Company agrees to buy the Dolly Varden Mine and Railway provided the charter is renewed (Chap. 7)
January 1919	R.W. Macintyre inspects the Dolly Varden Mines Railway and reports that the railway was not completed in time to comply with the charter (Chap. 7)

February 1919	Taylor petitions the Provincial Government to save his company and other creditors from being deprived of more than $467,000.00 by the Dolly Varden Company (Chap. 8)
February 12, 1919	A select committee of the Provincial Legislature is convened to hear evidence and make recommendations on the Dolly Varden charter and petitions (Chap. 8)
March 1919	The Taylor Engineering Company acquires the Dolly Varden Mine and Railway under the provisions of the Dolly Varden Mines Railway Amendment Act (Chap. 8)
May 1919	The Taylor Mining Company is capitalized for over a million dollars by its board of directors largely made up of former creditors of the Taylor Engineering Company (Chap. 9)
June 4, 1919	The unpaid workers are paid at last from the moneys raised in the capitalization of the Taylor Mining Company (Chap. 9)
July 1919	William Rae inspects the Dolly Varden Mines Railway for the Province of British Columbia, condemns the two 4-wheel cabooses and requires a number of improvements. (Chap. 9, App. A)

Ore Shipping on the Dolly Varden Mines Railway – July 1919 to November 1920

July 1919	The first ore is shipped over the Dolly Varden Mines Railway (Chap. 9)
September 1919	The Dolly Varden Mine shipped over 150 tons of ore daily over the Dolly Varden Mines Railway in September of 1919 (Chap. 9)
December 16, 1919	The last ore for the 1919 season is shipped on the Dolly Varden Mines Railway (Chap. 9)
December 20, 1919	Dog sleds begin hauling ore from the Dolly Varden mine over the right-of-way of the Dolly Varden Mines Railway (Chap. 9)
January 1920	The *Portland Canal Miner* announces the discovery of ore that is 75 to 80 percent pure silver at the Dolly Varden Mine (Chap. 10)
February 25, 1920	The miners and railway workers go on strike for higher wages (Chap. 10)
1920	To qualify for an operating certificate, half the Dolly Varden Mines Railway had to be re-ballasted, large ore bunkers had to be constructed at the tidewater and at the mine, and a high line had to be constructed for running ore cars out over the bunkers for dumping the ore (Chap. 10)
July 1920	Over 5,000 tons of ore shipped over the Dolly Varden Railway (Chap. 10)
August 1920	Dolly Varden Mines Railway Locomotive #5 becomes a runaway and jumps the track on the 5½ percent grade (Chap. 10)
August 1920	A. McPhee's inspection of the accident involving Locomotive #5 reveals serious deficiencies in operational procedures, construction, and maintenance on the Dolly Varden Mines Railway (Chap. 10)
August 1920	The Taylor Mining Company lets contracts for the first mile of extension of the railway to the Wolf Mine (Chap. 10)
August 1920	A fierce 4-day rainstorm causes slides and washouts on the Dolly Varden Mines Railway but it is back in service within a few days (Chap. 10)
November 1920	Al Falconer, a citizen of Alice Arm, gets a contract to haul ore by horse-drawn sleds during the winter and begins to plank the bridges in preparation for this work (Chap. 10)

The Death of the Dream Mine – December 1920 to Autumn 1922

December 1920	The Taylor Mining Company begins to lay off some of its 250 employees and slows production for the winter (Chap. 10)
1920	The price of silver drops to less than $1.00 per ounce and the Taylor Mining Company searches for more capital (Chap. 10)
April 1921	George Wingfield of Goldfield Consolidated Mines Company sues for the $150,000.00 plus interest owed him by the Taylor Mining Company and receives a mortgage on the mines and a lease on the Dolly Varden Mines Railway for 999 years (Chap. 10)
July 4, 1921	Official opening of the Dolly Varden Mines Railway under operating certificate #404 of the B.C. Department of Railways (App. A)
July 30, 1921	Rail shipments of ore begin to roll down to Alice Arm (Chap. 10)
August 13, 1921	The Taylor Company announces that ore shipments will cease at the end of the next week; shipments cease and only speeders and gas-mechanical locomotives run on the railway after this date (Chap. 10)
November 1921	A rainstorm closes the Dolly Varden Mines Railway completely, even to speeder traffic (Chap. 10)
July 1, 1922	The government of British Columbia repairs the railway so light rail traffic can traverse it (Chap. 10)
Autumn of 1922	George Wingfield and his associates take over the assets of the Dolly Varden Mines Railway and lease some of the Dolly Varden Railway equipment to the Granby Company for use in logging (Chap. 11)

The Logging Years on the Dolly Varden Mines Railway – Spring 1923 to December 1927

Spring 1923	The Abbotsford Lumber, Mining, and Development Company builds a narrow gauge railway on the Kitsault Flats and logs an area between the mouths of the Kitsault and Illiance Rivers (Chap. 12)
December 1923	The Granby Consolidated Mining, Smelting and Power Company announces logging operations along the Dolly Varden Mines Railway and on the Kitsault Flats in 1924 (Chap. 12)
May 6, 1924	Dolly Varden Railway Locomotive #6, a Climax, pulls the first narrow gauge log train to Alice Arm from Camp 2 (Chap. 12)
July 1924	An average of 20 carloads of logs per day are shipped over the lower portion of the Dolly Varden Mines Railway under its lease to the Granby Company (Chap. 12)
April 1925	Premier Oliver states that the government is willing to help any mining company that wishes to use the Dolly Varden Mines Railway (Chap. 11)
May 1, 1926	The Dolly Varden Mines Railway is opened for the season after Toric Mines, Limited placed it in operating condition with the government paying the costs (Chap. 11)
September 15, 1924	All logging in the Alice Arm area is suspended due to a downturn in the lumber market (Chap. 12)
May 1926	F.L. Housley takes over the management of logging for the Granby Company and begins construction of a bridge across the northeast fork of the Kitsault River and laying of more track (Chap. 12)
October 1926	Log carrying hulks arrive to pick up the first shipments of lumber arriving across the now completed 100-foot bridge across the Kitsault River (Chap. 12)
December 1927	The final shipment of lumber leaves the Alice Arm area and the railway logging years are over (Chap. 12)

The Dolly Varden Mines 1929 to 1980

November 1929	The Britannia Mining and Smelting Company options the Dolly Varden properties and does some development work. The option is dropped due to a drop in silver prices due to the Great Depression (Chap. 13)
1935	T.W. Falconer leases the Dolly Varden Mine and begins small ore shipments via speeder which lasted through 1940 (Chap. 13)
1937	The Dolly Varden Mines Railway Equipment is cut up by a scrapper (Chap. 13)
1978-80	Some development work is carried out at the Dolly Varden Mine and associated properties (Chap. 13)
1979	Ore reserves at the Dolly Varden Mine are reported at 47,443 tons on the basis of diamond drilling reports from the 1920s (Chap. 13)
1980 - 1985	Dolly Varden Minerals, Inc. acquires and holds the Dolly Varden Mines (Chap. 13)

Numbered Notes

Numbered notes are in parentheses in the text and have been kept to a minimum in the interests of readability. Much of the information in this book was gleaned from the "Dolly Varden Papers" now filed as part of the Clerk of the House's Papers for 1919 in the Provincial Legislative Library at Victoria, B.C., Canada. The "Dolly Varden Papers" consist of several hundred pages of testimony typed by government reporters, with supporting correspondence from the files of the Taylor Engineering Company, letters and cables from Joseph Hubbard, and other documentary evidence taken by the select committee of the British Columbia Legislature in the spring of 1919. All quotations attributed to Alfred Taylor and Joseph Hubbard were taken directly from the sworn evidence given by them at the Dolly Varden hearing. Other quoted materials are from sources cited in the text or in these notes.

Besides the numbered notes, there follows a reference list of other printed materials consulted in the preparation of this book. All newspapers listed are available on microfilm in the Provincial Archives at Victoria, B.C.

Chapter 1

1.1 Mining and Scientific Press, May 1, 1920 p. 631 (B.C. Legislative Library)

1.2 Provincial Archives of British Columbia, Report of the Minister of Mines for 1913 pp. K79-K83, British Columbia Sessional Papers, 1913.

Chapter 2

2.1 Provincial Archives of British Columbia, Report of the Minister of Mines for 1916. British Columbia Sessional Papers, 1916.

Chapter 4

4.1 Victoria B.C., B.C. Legislative Library, letter from L. W. Patmore to Premier Oliver of January 13, 1917. Clerk of the House Papers for 1919.

Chapter 5

5.1 Victoria B.C., B.C. Legislative Library, "Dolly Varden Papers," p. H-19, Clerk of the House Papers for 1919.

5.2 Victoria B.C., B.C. Legislative Library, "Dolly Varden Papers," p. L-25, Clerk of the House Papers for 1919.

5.3 Victoria B.C., B.C. Legislative Library, letter from A.J. Taylor to F.M. Sylvester, Jan. 29, 1918. "Dolly Varden Papers," Clerk of the House Papers for 1919.

Chapter 6

6.1 Victoria B.C., B.C. Legislative Library, letter to Taylor Engineering Company from J.D. Hubbard, May 14, 1918, "Dolly Varden Papers," Clerk of the House Papers for 1919.

6.2 Victoria B.C., B.C. Legislative Library, letter from Simon Maloney to A.J. Taylor, June 28, 1918. Clerk of the House Papers for 1919.

6.3 Victoria B.C., B.C. Legislative Library, testimony of Alfred Taylor, "Dolly Varden Papers," p. L-36, Clerk of the House Papers for 1919.

6.4 Victoria B.C., B.C. Legislative Library, nightletter from A.J. Taylor to J.D. Hubbard, July 31, 1918 "Dolly Varden Papers," Clerk of the House Papers for 1919.

Chapter 7

7.1 Victoria B.C., B.C. Legislative Library, "Dolly Varden Papers," p. L-36, Clerk of the House Papers for 1919.

7.2 Ibid., p. L-37.

7.3 Ibid., p. L-37.

7.4 Ibid., pp. L-39, L-40.

7.5 Ibid., p. 40.

7.6 Ibid., p. 41.

7.7 Ibid., p. L-41.

7.8 Victoria, B. C., Provincial Archives, railway inspection report by the British Columbia Department of Railways.

7.9 Victoria B.C., B.C. Legislative Library, nightletter from J.D. Hubbard to H.C. Chiene, Oct. 11, 1918, "Dolly Varden Papers," Clerk of the House Papers for 1919.

7.10 Victoria, B.C., Provincial Archives, letter from Premier John Oliver to Chief Engineer Proctor of the Department of Railways, dated Dec. 4, 1918, B.C. Railway Department files.

Chapter 8

8.1 Victoria B.C., B.C. Legislative Library, letter and draft legislation submitted to Premier John Oliver by L. W. Patmore on Feb. 3, 1918, "Dolly Varden Papers," Clerk of the House Papers for 1919.

8.2 Victoria B.C., B.C. Legislative Library, letter from Alfred Taylor to Charles Hamill, Feb. 3, 1919, "Dolly Varden Papers," Clerk of the House Papers for 1919.

8.3 Victoria B.C., B.C. Legislative Library, letter from Alfred Taylor to Premier John Oliver, "Dolly Varden Papers," Clerk of the House Papers for 1919.

8.4 Victoria B.C., B.C. Legislative Library, Taylor's petition to Premier Oliver, "Dolly Varden Papers," Clerk of the House Papers for 1919.

8.5 Victoria B.C., B.C. Legislative Library, "Dolly Varden Papers," pp. H-3, H-4, Clerk of the House Papers for 1919.

8.6 Victoria B.C., B.C. Legislative Library, "Dolly Varden Papers," p. 72 (from cross-examination of A. Taylor), Clerk of the House Papers for 1919.

8.7 Victoria B.C., B.C. Legislative Library, cross-examination of J.D. Hubbard by Mr. Mayers for the Taylor Engineering Company, "Dolly Varden Papers," p. 221, Clerk of the House Papers for 1919.

8.8 *Vancouver Daily World*, Mar. 24, 1919, report on the findings of the select committee.

8.9 Statutes of British Columbia for 1919, Dolly Varden Mines Railway Amendment Act.

8.10 *Vancouver Daily World*, Mar. 27, 1919.

Chapter 9

9.1 *Mining and Scientific Press*, July 5, 1919, p 27.

9.2 Victoria, B.C., Provincial Archives, railway inspection report by the British Columbia Department of Railways dated July 11, 1919.

Chapter 10

10.1 *Vancouver Daily World*, Apr. 17, 1920.

10.2 Victoria, B.C., Provincial Archives.

10.3 *British Columbia Gazette*, Order in Council 347/21 of Mar. 15, 1921.

Other References

NEWSPAPERS

Mining references from the *Alice Arm and Anyox Herald*

1920 Nov. 4
1921 Oct. 15; Nov. 5, 6; Dec. 17.
1922 Feb. 11; Mar. 18; Apr. 1; May 6, 27; Jun 3; Jul. 1, 8; Oct. 14; Dec. 23.
1923 Jan. 20; May 31; Jun. 14; Aug. 9, 23, 30.
1924 Jan. 5; May 31; Jun 14; Aug. 9, 23, 30.
1925 Jan. 10, 31, Mar. 7, 14, 21; Apr. 4, 11; May 23; Jun 6, 13; Sep. 14, 26; Oct. 17; Dec. 5.
1926 Jan. 9, 16; Feb. 13; Mar. 13, 20; Apr. 24; May 1; Jul 2, Sep. 11, 18, Oct. 16; Nov. 26; Dec. 11.
1927 Mar. 5; Jun. 4; Oct. 8, 15.
1928 Feb. 4; Feb. 11; Jun. 9, 29; Aug. 3, 17.
1929 Mar. 30; Apr. 6, 20; May 25; Jun. 8; Aug. 23; Sep. 6; Nov. 23, 30; Dec. 7.
1930 Mar. 1; Apr. 17, 26; May 3; Aug 22; Sep 20, Oct. 4; Nov. 1.
1931 Mar. 14; Jun. 6.
1932 May 28; Sep. 24; Oct. 15.
1933 Apr. 1; Jun 17; Nov. 11.
1934 Jan. 20; Feb. 3; Mar. 24; Apr. 7, 14; May 5; Jun 9; Aug. 31, Sept. 8, 22; Oct. 6, 27; Nov. 3
1935 Mar. 2, 9. (The issue dated Mar. 9 was the final issue and was actually published on May 4, 1935)

Lumbering references from the *Alice Arm and Anyox Herald*

1922 Sept. 23; Oct. 7; Nov. 11; Dec. 9, 23.
1923 Feb. 10; May 26; Jul. 14, 28; Aug. 4; Sep. 8, 29; Oct. 13; Dec. 1, 15, 22, 29.
1924 Feb. 9, 23; Mar. 1, 8, 15, 22, 29; Apr. 2, 19; May 17, 23, 31; Jul 5, 12; Aug. 16, 30; Sep, 20; Oct. 10, 17, 24; Nov. 15; Dec. 20
1925 Feb. 7, 28; Apr. 4; May 2, 10; Aug. 10; Sep. 19.
1926 Jan. 23, Mar. 20, May 8, 29; Aug. 13; Oct. 2, 9; Nov. 20.
1927 Jan. 8; Feb. 12, Apr. 9, 30; May 7, 14, 21; Jun. 4; Aug. 26, Oct. 20; Dec. 31.
1928 May 12; Jun. 2, 22; Nov. 29.
1930 Nov. 29.
1931 Nov. 28.

The *Cassiar News* (Stewart, B. C.)

1919 Jun. 13.

The *Portland Canal News* (Stewart, B. C.)

1919 Oct. 10, 17; Dec 20.
1920 Jan. 10, 17; Jul. 3.
1927 Jun. 3; Sep. 30; Oct. 28.

The *Vancouver Province* (Vancouver, B. C.)

1919 Jun. 4, 7.
1920 Apr. 16; May 1.
1929 Sep. 1.
1932 Jul. 3.

The *Vancouver Daily World* (Vancouver, B.C.)

1919 Mar. 24, 25, 27; Jun. 28; Jul. 19; Aug. 2, 9; Sep. 6, 13, 27; Oct. 4, 31; Nov. 1; Dec. 20, 27.
1920 Jan. 17; Feb. 14, 21; Mar. 6, 27; Apr. 3, 10, 17; Jul. 3, 5, 17, 24; Aug. 14, 28.

The *Victoria Daily Colonist* (Victoria, B. C.)

1920 May 1.
1928 Aug. 5.

(All of these newspapers are on microfilm in the B. C. Provincial Archives at Victoria, B.C. Canada)

OTHER PERIODICALS

The *British Columbia Lumberman* (Vancouver, B. C.)

1931 January issue
1932 January issue

The *Canadian Mining Journal* (Toronto, Ontario)

1919 Feb. 26; Mar. 12; Jun. 18.
1920 Jul. 9, 16, 23; Sep. 17, 24; Nov. 5, 11.
1921 May 27; Jun. 10; Jul. 29; Sep. 2; Nov. 11; Dec. 30.
1922 Jan. 13; Feb. 24; May 19; Jun. 2, 9, 23, 30; Jul. 7; Sep. 15; Oct. 20.

Mining and Scientific Press

1919 Jul. 5 (p. 27), 12 (p. 69); Sep. 13 (p. 379), 20 (417), 27 (454); Oct. 11 (p. 543); Nov. 8 (p. 687)
1920 Jan. 1 (p. 27), 24 (133); Feb. 28 (pp. 323-324); Mar. 27 (466); Apr. 10 (545); May 15 (p. 725); Jun. 5 (pp 845-846); Jun. 12 (p. 879); Jun. 19 (919), 26 (p. 955); Aug. 19 (pp. 207-208).
1921 May 14 (p. 694); Jun. 4 (p. 793), 18 (p. 858).

OTHER DOCUMENTS

Annual Report of the Minister of Mines 1910 to 1958. (In the *British Columbia Sessional Papers*, B.C. Provincial Archives, Victoria, B.C., Canada)

British Columbia Railway Department Inspection Reports (on file in the B.C. Provincial Archives, Victoria, B.C. Canada)

The *Dolly Varden Papers* (in the Provincial Legislative Library, Victoria B.C. Canada)

Reports of the Canadian Geological Survey (on file in the B.C. Provincial Legislative Library, Victoria, B.C. Canada)

Photo Credits

(Placement code: T – top; M – middle; B – bottom; LT – left top; RB – right bottom, etc.)

Walter C. Casler: pp. 85, 150T
Marilyn Blaisdell: p. 102
B.C. Forest Service via Elwood White: p. 131B
B.C. Forest Service photo by J.B. Scott: p. 132
B.C. Provincial Archives: pp. 151M, 163, 172B (map)
B.C. Sessional Papers 1914: pp 172T, 173B (map)
Ronald Evindsen: pp. 6, 8, 9, 10M, 10B, 11T, 11B, 13B, 14, 17T, 17B, 18, 19, 36T, 36B, 37, 42T, 47, 55, 73, 74, 94, 108B, 140, 143LT, 143LB, 143RT, 143RB, 149
Alma Evindsen Graham: pp. 5T, 5B, 7, 10T, 12, 13T, 51, 71-72, 84B, 10T, 127, 135, 136, 139
Granby News, Vol. 3 #5, May 1919: 173B
Granby News, Vol. 1, #6, November 1917: 174T, 174B
Tom Gray, courtesy of Douglas Richter: p. 175T
Patrick O. Hind: pp. 147, 164, 165T, 165B
J.E. Lewis: pp. 150B, 153, 154T, 154M, 154B, 155, 164T, 166T, 167T, 167B, 168T, 168B, 169T, 169B, 171
Darryl Muralt: pp. 19, 21, 40, 45, 48, 49, 54, 80, 81T, 82T, 82B, 83-84, 124, 131T, 141, 142, 151B, 173T
Vancouver Public Library: pp. 1, 15, 27, 34, 41, 42B, 58, 59, 64, 66, 77, 78, 79, 81B, 86, 87, 88, 89, 90T, 90B, 91, 92, 93, 95, 97, 99, 100, 101, 103, 104, 105, 106, 107, 109, 111, 112, 113, 114, 115, 118, 121, 123, 125, 126, 129, 137, 144, 145

Index

Page references in bold face type indicate item appears in a photo, map or illustration on that page. The abbreviation DVMR stands for Dolly Varden Mines Railway.

This copy of Steel Rails & Silver Dreams was printed on 135 gsm matte coated stock in China. The typeface is Goudy Old Style.

Alice Arm
and the
Kitsault Valley
1923
BRITISH COLUMBIA, CANADA
Lake
Trout lakes
Trout Creek
Ridge Line
WOLF MINE
Black Bear Creek
DOLLY
VARDEN
HILL
DOLLY
VARDEN
MINE
17
CAMP 8
Varden
Glacier
UPPER
CANYON
16
Homestead
Creek
15
5 1/4% GRADE
Evindsen Creek
Kinskutch Lake
Ridge Line
14
East Creek
Ice Fields
West Creek
13
0
1
2
3
Miles
Mt. Evindsen
Ridge
Line
RAILWAY
12
MINES
11
VARDEN
Kitsault River
10
DOLLY
Ice Fields
Ridge Line